New Guinea Vegetation

New Guinea Vegetation

K. Paijmans, Editor

ELSEVIER SCIENTIFIC PUBLISHING COMPANY
Amsterdam – Oxford – New York   1976

Published in co-edition with
The Australian National University Press,
Canberra

Distribution of this book is being handled by the following publishers

    for the U.S.A. and Canada

Elsevier North-Holland Inc.
52, Vanderbilt Avenue
New York, N.Y. 10017

    for Australia, New Zealand and Papua New Guinea

The Australian National University Press, Canberra

    for all remaining areas

Elsevier Scientific Publishing Company,
335 Jan van Galenstraat
P.O. Box 211, Amsterdam, The Netherlands

ISBN 0-444-99827-6

© CSIRO 1976

This book is copyright. Apart from any fair dealing for the purpose of
private study, research, criticism, or review, as permitted under the
Copyright Act, no part may be reproduced by any process without
written permission. Inquiries should be made to the publisher.

First published in Australia, 1976

Printed in Hong Kong for the Australian National University Press, Canberra

Preface

Resources reconnaissance surveys in Papua New Guinea by the Division of Land Use Research, CSIRO, began in 1952, at the request of the Papua New Guinea Administration, which financed the whole project. Following fifteen regional surveys, it was decided to produce a book on the vegetation based on the survey information. As it seemed desirable to widen the scope of the book, two scientists outside CSIRO were invited to contribute. Thus the book came to consist of three parts: in Part I Max van Balgooy discusses the origin and distribution patterns of the flora; Part II gives a comprehensive account of the main vegetation types; and in Part III Jocelyn Powell describes the role of the vegetation in the life of the native population. Part I deals with the whole New Guinea region, while Parts II and III are concerned mainly with Papua New Guinea.

The individual parts can be read separately, although cross-references are made where appropriate. The views of the authors do not always coincide, and the vegetation history in particular, which is based partly on conjecture, has given rise to some differences of opinion. The authors have tried to avoid unnecessary technical terms so as not to discourage readers who are unfamiliar with the jargon of a particular discipline. All references cited in the parts have been brought together in a single bibliography at the end of the book. With very few exceptions measurements are in metric units. Botanical names have, as far as practicable, been brought up to date and may be different from those used in the literature cited, but well known synonyms are added in brackets. Because of many uncertainties and changes in nomenclature it was not possible to provide all botanical names with the names of the original authors and it was therefore decided not to quote any authorities. Thanks are due to Dr. T.G. Hartley of the Herbarium Australiense, Canberra, who has checked the botanical names used.

Plates 1-43, 46, 48 and 49 are from the CSIRO Papua New Guinea surveys collection, and 44, 47 and 50-3 are Dr Powell's. The aerial photographs in Plates 16, 25, 33, 40 and 42 are Crown Copyright reserved. Their publication has been made possible by courtesy of the Director, Division of National Mapping, Department of Minerals and Energy, Canberra. The figures were prepared by the Drawing Office of the CSIRO Division of Land Use Research, Canberra.

The island of New Guinea, lying wholly within the tropics, has environments ranging from coastal swamps to snow-capped mountain peaks. Our knowledge of its flora is still incomplete and patchy; in some areas it is known reasonably well, but in other areas, which remain unexplored, the flora has yet to be investigated. Consequently there is much scope for further study. The vegetation of New Guinea is relatively unspoilt in comparison with many other tropical areas, and many natural habitats can still be preserved, a challenging task for a newly independent country.

K. Paijmans

Contents

Preface   v
Tables   xi
Figures   xiii
Plates   xv
Note on the use of 'New Guinea'   xvii

Part I   **Phytogeography**   M.M.J. van Balgooy

Introduction   1
Floristics of New Guinea   5
    Families   5
    Genera   8
    Species   16
Origin of the New Guinea flora   16

Part II   **Vegetation**   K. Paijmans

Introduction   23
Major environments and structural forms   24
Method of description   25
The beach ridges and flats   27
    Vegetation and habitat   27
    Vegetation types   28
    Man-made vegetation   30
The saline and brackish swamp   31
    Mangroves and their habitat   31
    Types of mangrove vegetation   32
    Man-made vegetation   36
    Resources   36
The lowland freshwater swamps   37
    Swamp vegetation and its habitat   37
    Types of swamp vegetation   37
    Plant succession   48
    Resources   49

The lowland alluvial plains and fans   49
    Vegetation and habitat   49
    The forest   49
    Environmental controls   51
    Other vegetation types   52
    Man-made vegetation   59
    Plant succession   59
    Resources   62
The foothills and mountains below 1000 m   64
    Vegetation and habitat   64
    The forest   64
    Environmental controls   65
    Predominance of certain trees   68
    Quantitative data on forest structure and floristics   70
    Other vegetation types   74
    Origin and status of eucalypt savanna and lowland grassland   79
    Man-made vegetation   80
    Plant succession   81
    Resources   84
The lower montane zone   84
    Vegetation and habitat   84
    The forest   85
    Environmental controls   86
    Predominance of certain trees   88
    Other vegetation types   91
    Origin and status of the high-mountain grasslands and tree fern savanna   94
    Man-made vegetation   96
    Resources   97
The upper montane zone   97
    Vegetation and habitat   97
Vegetation history   101
The vegetation of Irian Jaya   104

Part III **Ethnobotany**   J.M. Powell

Introduction   106
Staple and supplementary crops   107
    The coastal and lowland swamp dwellers   112
    The lowland shifting agriculturalists   118
    The highland margins   126
    The highland agriculturalists   126
Other useful plants   134
    Narcotics, stimulants and intoxicants   134
    Medicinal plants   135
    Ritual and magic   147
    Art   150
    Tools and weapons   152
    Hunting and fishing   155
    Canoes and rafts   157
    House building   160
    Food preparation, containers and vessels   166
    Cordage, bark cloth and other textiles   168
    Clothing and personal ornamentation   170
Discussion   174
    Subsistence agriculture: the plant base   175
    Former vegetation and environment: a hypothetical reconstruction   176
    Archaeological and palynological evidence for agriculture   179
    Other useful plants and products   181
    Population: the nutritional base   182
    The present situation   183

Bibliography   185
Index of Botanical Names   201
General Index   210

Tables

1.1 Families in West Malesia not known from New Guinea 6
1.2 Families in Australia not known from New Guinea 6
1.3 Families in New Guinea not known from West Malesia 7
1.4 Families in New Guinea not known from Australia 7
1.5 Number of genera with indigenous species in New Guinea per distribution type 11
1.6 Number of genera per distribution category 13
1.7 Number of montane genera per distribution type 15
1.8 Number of montane genera per distribution category 15
1.9 Annotated list of mountain genera 19

2.1 Major environments and vegetation types 26
2.2 Numbers of individuals and species of trees with a diameter breast height of 10 cm and over in mixed tropical rain forest 71

3.1 Plants used as food 108
3.2 Chemical composition of staple crops 115
3.3 Chemical composition of some supplementary crops 116
3.4 Chemical composition of coconuts 123
3.5 Plants used in the treatment of cuts and wounds 136
3.6 Plants used in the treatment of burns 137
3.7 Plants used in the treatment of sores 137
3.8 Plants used to relieve headaches, general body pains and swellings 140
3.9 Plants used to relieve toothache and other mouth infections 141
3.10 Plants used in the treatment of fevers, including malaria 142
3.11 Plants used in the treatment of coughs, colds and sore throats 143
3.12 Plants used in the treatment of dysentery, diarrhoea and stomachaches 144

- 3.13 Plants associated with the control of fertility and childbirth 147
- 3.14 Plants used in rituals and magic 148
- 3.15 Plants used to make tools and weapons 152
- 3.16 Plants used in the construction of canoes and rafts 158
- 3.17 Plants used in house, shelter and fence building and decoration 162
- 3.18 Plants used as ropes in construction of houses, shelters, fences, etc. 165
- 3.19 Leaves used for lining cooking ovens and wrapping food for cooking 167
- 3.20 Plants used for making string and bark cloth 169
- 3.21 Leaves and flowers used in everyday and ceremonial dress 172
- 3.22 Other species used in personal adornment 173

Figures

1.1 Demarcation lines around New Guinea   1
1.2 Relation between number of genera and area   5
1.3 Distribution of *Connarus, Coriaria, Batis*   9
1.4 Distribution of *Rhododendron, Pittosporum, Balanophora*   10
1.5 Distribution of *Kopsia, Dendromyza, Archidendron*   10
1.6 Distribution of *Sericolea, Hibbertia, Eupomatia*   12
1.7 Distribution of *Bikkia, Quintinia, Nothofagus*   12
2.1 Topographic map of New Guinea   23
2.2 Major environments of Papua New Guinea   24
2.3 Number of trees per girth class of 1 foot on 0.8 ha   72
2.4 Number of species per unit area, average of four 0.8 ha sample plots   73
3.1 Locations of specific studies consulted   107

Plates

1 *Ipomoea pes-caprae* and other creeping herbs, sedges and grasses 28
2 Planted coconut replacing natural vegetation 30
3 Cut-off river section being colonised by *Sonneratia* 32
4 Mangrove forest lining saline tidal creek 33
5 Mangrove forest with ground layer of *Acrostichum aureum* 33
6 Woodland of *Avicennia* 35
7 Dense nipa palm vegetation in estuary 36
8 Herbaceous swamp vegetation 38
9 Sepik River area 39
10 Swamp grass lining flooded river 40
11 *Saccharum robustum* 40
12 Seasonally dry *Pseudoraphis* grassland 41
13 Sago palm woodland 43
14 *Pandanus* swamp woodland 44
15 Swamp forest in fluctuating back swamp 45
16 *Campnosperma* swamp forest 45
17 *Terminalia brassii* swamp forest 47
18 *Melaleuca* swamp forest 47
19 Interior of tall forest on alluvium 50
20 Tall mixed savanna on undulating terrain 53
21 *Melaleuca* savanna on seasonally inundated plain 54
22 Tall *Saccharum spontaneum-Imperata cylindrica* on plain 56
23 Sedge-grassland on seasonally inundated flat 58
24 Pioneering vegetation in inner river curve 60
25 Pioneering vegetation from the air 60
26 *Pterocarpus indicus* being felled 63
27 *Araucaria hunsteinii* forest 63
28 Dense thin-stemmed forest on low hills 67
29 Buttressed trunk of old *Casuarina papuana* tree 69
30 Fire-disclimax *Themeda australis* grassland 75
31 Eucalypt savanna 77
32 Eucalypt savanna and forest on foothills 77
33 Areas of grassland reverting to forest 78
34 Pioneering vegetation on lava flow 80

35 Interior of lower montane forest 82
36 'Elfin woodland' 83
37 Interior of *Nothofagus* forest 85
38 Grassland, tree fern savanna and lower montane forest 88
39 Tree fern savanna of *Cyathea* 89
40 Grassland, tree fern savanna and lower montane forest from the air 90
41 Lower and upper montane forest 93
42 'Fern-leaf' pattern of eucalypt savanna and forest 93
43 Garden with taro and bananas 98
44 Woman making sago 113
45 Taro and sweet potato garden 121
46 Sweet potato garden 127
47 Sugarcane, lima bean and sweet potato garden 128
48 Trunk of *Calophyyllum* tree being hollowed out for canoe 157
49 Lowland village in Western District 161
50 Typical highland house 161
51 *Pandanus* fibre and rope in Tari market 170
52 Netted string aprons 171
53 Women's clothes in the highlands 171

Note on the use of 'New Guinea'

In accordance with normal scientific practice 'New Guinea' has been used throughout this book to mean the whole island of which Papua New Guinea and Irian Jaya are political divisions. When Papua New Guinea is used, it refers specifically to the political entity that bears that name. The 'highlands' refer to the mountains and intermontane valleys of the central cordillera of New Guinea, while the 'Highlands' refer to the administrative districts in Papua New Guinea.

PART I

Phytogeography

M.M.J. van Balgooy

Introduction

The island of New Guinea occupies a phytogeographically strategic position between Asia and West Malesia (Malesia in the botanical sense comprises Indonesia, Malaysia, the Philippines, Portuguese Timor and New Guinea; *vide* van Steenis 1950) on the one hand and Australia and the Pacific on the other. The biogeographic importance of New Guinea, and the south-west Pacific generally, is apparent from the numerous contributions that have been written concerning various aspects of its flora and fauna. It is further illustrated by the number of biogeographic demarcation lines that have been drawn in this part of the world (Fig. 1.1). The most famous among these are the 'Wallace line' and the Torres Strait boundary.

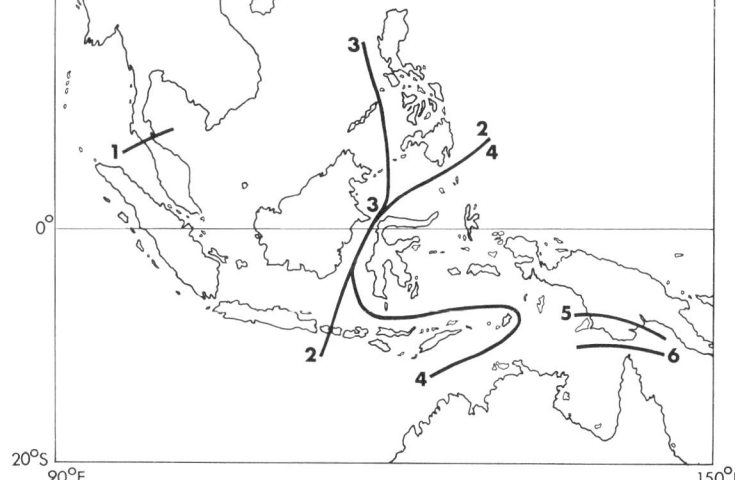

Figure 1.1
Some biogeographical demarcation lines around New Guinea.
1. Ridley's line
2. Wallace's line
3. line of Merrill and Dickerson
4. Zollinger's line
5. Good's line
6. Torres Strait Boundary.
Data from van Steenis (1950), Good (1963) and Keng (1970).

As is well known, the distributions of flora and fauna groups in this area have given rise to much controversy. Phytogeographers place New Guinea in the Indo-Malesian Floral Region (see for example van Steenis 1950), whereas zoogeographers, mainly those basing their case on the distribution of mammals and birds, place New Guinea in the Australian Region (see for example Darlington 1957). Students of invertebrate groups tend to agree with the phytogeographers' standpoint (see for example Gressitt 1961).

Thanks are due to Professor C.G.G.J. van Steenis for critically reading the first draft, to Professor C. Kalkman and Dr C.E. Ridsdale for general suggestions, and to Dr P. van Royen for comments on the list of montane genera.

A discussion of the phytogeography of Papua New Guinea is meaningless if the rest of the island is left out of consideration. Therefore this part will treat the flora of the whole of New Guinea, including the adjacent islands: Aru, Kai, Raja Ampat, Yapen/Biak, Bismarcks, Solomons, Louisiades and the d'Entrecasteaux Islands.

This part deals only with the Phanerogams (Spermatophytes), i.e. the Angiosperms (flowering plants) and Gymnosperms. Therefore the conclusions reached here can be applied only with some caution to other groups of plants such as the Bryophytes and the Ferns. The distribution of plants carried about and cultivated by man is a field of study in its own right and is treated by Powell in Part III of this book. In this part only the indigenous plants (i.e. those occurring naturally) will be dealt with.

Phytogeographic studies invariably have to start with the compilation of facts. In other words, we have to find out what plants occur in New Guinea and how they are distributed over the face of the earth. Every taxon or systematic unit of plants (species, genus, family) has a unique distribution, but closer examination reveals certain patterns. These patterns and details of the presence or absence of taxa give indications about evolution and past tracks of migration and barriers, although conclusions are often tentative and speculative.

For the compilation of distribution data of the New Guinea flora I have consulted the series 'Beiträge zur Flora Papuasiens' in *Botanische Jahrbücher*, mainly concerning the former German territory in north-east New Guinea, the publications in *Nova Guinea* concerning the former Dutch New Guinea, the series 'Plantae Papuanae Archboldianae' by Merrill and Perry in the *Journal of the Arnold Arboretum*, and numerous other articles in several journals by various authors. Helpful sources in the compilation included Willis/Airy Shaw (1973), van Royen (1959) and Coode (1969b). These data were supplemented from unpublished herbarium records. Sources of information outside New Guinea were, for Malesia: unpublished data by van Steenis; for Australia: Burbidge (1963), Chippendale (1972) and information provided by Hyland and colleagues of the Atherton Forestry and Timber Bureau; for the Pacific: van Balgooy (1971). A complete list of genera would take too much space. The reader is given only a few tables in which the facts are expressed in figures, the only way to allow evaluation.

Several authors have discussed phytogeographical aspects of the flora of New Guinea and many have attempted to unravel its origin. Some authors have treated a limited group of taxa, e.g. Mattfeld (1929) the Compositae, Diels (1913) the Annonaceae, Smith (1943, 1945) the Winteraceae, van Steenis (1953, 1971) *Nothofagus*. Others have dealt with areas of particular interest, for example Gibbs (1917) Arfak in western New Guinea, Diels (1930) the mountain flora, Whitmore (1969) the Solomons, Heyligers (1972c) the Port Moresby area of New Guinea. Some have taken into account the whole Phanerogam flora: Warburg (1891), Lam (1934), van Steenis (1954), Good (1960), Walker (1972, especially the contribution by Hoogland). Relevant papers not dealing directly with New Guinea include Kalkman (1955), Burbidge (1960), van Steenis (1962a),

Thorne (1963), Smith (1970), van Balgooy (1971), Raven and Axelrod (1972), Whitmore (1973). These have been helpful in preparing this part.

Before attempting to draw conclusions from the facts collected certain assumptions have to be made:

1. Every taxon is believed to be of monotypic origin, i.e. it started in a locality of limited size and from there expanded to surrounding areas.

2. In general, the higher the rank of a taxon, the older it will be. A species will be evolutionarily younger than the genus to which it belongs, and the genus will be younger than the family of which it is a member.

3. The presence of many species in common between two adjacent areas is considered a sign of recent exchange. On the other hand, if two areas are occupied by many different taxa, especially of high rank (genera and families), it is assumed that the two areas have had no recent contact.

4. Although a taxon is considered to have originated in a locality of limited size, it may in time radiate from the original locality, thereby often developing new forms. The area where most related forms occur, in other words the centre of diversity, in general indicates the centre of origin of the taxon. The centre of a genus need not be the area where most of the species occur at present. The classical example is that of the tribe Stapelieae-Ascl. The greatest number of mostly closely allied species of this tribe occur in South Africa. In India there are fewer species, but morphological differentiation is greatest, and some primitive forms occur. Hence it is most likely that the taxon originated in India and that later strong secondary speciation occurred in South Africa.

5. Theoretically, given time and all other factors being equal, the distribution of a taxon is proportional to its age, but since all factors are rarely equal it is hazardous to assume that taxa of limited distribution are young, and that widely distributed ones are old. The older a taxon, the more chance it has had to expand, but its success depends on its dispersal ability, its ecological tolerance, and the presence or absence of biological and physical barriers. Local extinctions may have taken place so that the present distribution of the taxon may be chequered. A discontinuous or disjunct distribution is often indicative of a former wide distribution. In some cases the distribution may crumble to such an extent that only a small area remains (this is called a relic distribution).

6. If free migration has been hampered for any reason, evolution may have taken place in isolation and have given rise to the development of endemic taxa, i.e. taxa of a limited distribution. The presence of many endemics, especially those of high rank, indicates long isolation of the area. Good examples of such isolated land areas are Madagascar and New Caledonia.

7. Lastly, we assume that the ecological amplitude of a taxon is genetically fixed and does not change drastically in time. In other words, if all species of a genus are at present confined to tropical climatic conditions, the genus was not a temperate one in an earlier phase of its evolution.

Thus, the biogeographer has to work mainly with circumstantial evidence. Very little can be proven conclusively except the actual occurrence of a certain taxon in some locality as documented by a specimen in the herbarium or a record in the literature, again assuming that the identification is correct. Furthermore, the biogeographer has to base most of his speculations on present-day distribution although, as we have seen, the area of a taxon may have expanded or contracted in time. Paleontological evidence (fossils, including those of pollen) is something to go on, but unfortunately the tropics generally have a rather poor fossil record compared with temperate regions. Moreover, fossil data are at best fragmentary. If a contemporaneous species is found in, say, Pliocene deposits of New Guinea, all we know is that the species existed in New Guinea in the Pliocene; we do not know whether it was absent previously, or whether it subsequently became extinct and resettled later.

The presence of a species of *Araucaria* on Norfolk Island isolated from other *Araucaria*-bearing countries by some 900 km 'proves' to one biogeographer that Norfolk Island has had terrestrial contacts with these countries, whereas to another it 'proves' that *Araucaria* is capable of crossing large water gaps. Actually only experiments on the dispersal ability of *Araucaria* seeds can indicate which theory is more likely to be correct.

There are strong reasons to assume that most migrations of land plants require at least intermittent or insular land connections (van Steenis 1962a). Successful dispersal (i.e. dispersal followed by permanent establishment) over large distances — so-called long-distance dispersal — must be considered a rare phenomenon generally, but it cannot be ruled out and may be effective, for example, in the case of beach plants. Moreover, in certain cases the loss of dispersal ability after successful establishment, an idea elaborated by Carlquist (1965, 1967), is a possibility. To the biogeographer the evidence based on well fossilised groups with proven poor dispersal ability and a limited ecological tolerance will carry more weight than evidence from taxa without all these qualities.

To evaluate the phytogeographic position of New Guinea we should have a complete list of species growing naturally on the island. No such list is available (*Flora Malesiana,* edited by van Steenis, is far from being completed). I have shown before (van Balgooy 1971), that there is general agreement among taxonomists concerning the delimitation of genera, but opinions often differ considerably as regards species and families. It is logical that the distribution of a species is much less completely known than that of the genus. For this reason my discussion on the phytogeography of New Guinea will lean heavily on the evidence presented by the distribution of genera.

Since families are older than genera and these are older than species, the distributions of each of these sets of taxa reflect different episodes of the earth's history. *Batis argillicola,* for example, is a species occurring in saline claypans in northern Australia and New Guinea. The only other species of *Batis* is confined to the Neotropics (Fig. 1.3c). The distribution of the genus is thus of a transoceanic type and is to the biogeographer a different problem from that of the distribution of its component species.

Floristics of New Guinea

The flora of New Guinea is one of the richest in the world. Expressed in the number of genera per unit area it is quite comparable to other tropical rain forest-covered islands and archipelagoes. This is demonstrated in Fig. 1.2 where the relationship of number of genera to area is shown (see also Part II). Let us start to place New Guinea

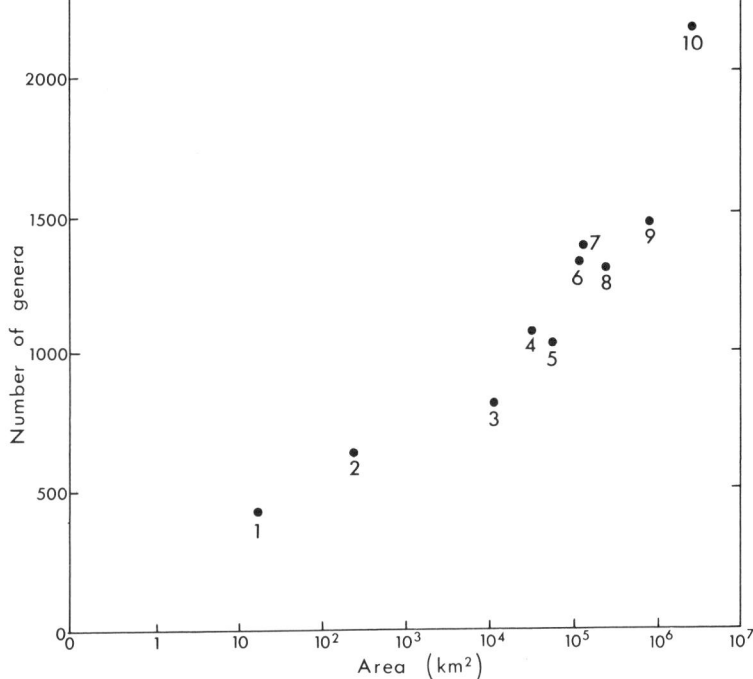

Figure 1.2
Number of genera plotted against log. area (km^2).
1. Barro Colorado, 445/15;
2. Penang, 633/270;
3. Jamaica, 810/11,400;
4. Taiwan, 1080/36,000;
5. Ceylon, 1044/62,500;
6. Java, 1320/125,000;
7. Malaya, 1407/132,000;
8. Philippines, 1308/290,000;
9. New Guinea, 1463/900,000;
10. Malesia, 2178/3,000,000.
Data from van Balgooy (1969), Beintema-Hietbrink (n.d., Ceylon), Keng (Malaya, 1970), Merrill (Philippines, 1926), van Steenis (Malesia, 1950).

in perspective by discussing the distribution of the taxa at various taxonomic levels, families, genera and species, and by comparisons with West Malesia and Australia.

Families

The total number of plant families present on the earth is subject to the system followed. In his *Syllabus* Engler (1954, 1964) accepts 355 Phanerogam families, eleven of which are Gymnosperms. If Willis/ Airy Shaw (1973) is followed the totals are 544 and thirteen respectively. Thorne (1968) accepts 411 Angiosperm families. It is clear that to make sensible comparisons with other parts of the world one system should be adopted. Under Engler's system New Guinea has 200 Phanerogam families; including families described lately this number would be 246. In Tables 1.1-4 all names appearing in Willis/Airy Shaw are taken up. The families not recognised by Engler are followed in parentheses by the name of the family from which they have been segregated, e.g. Carlemanniaceae (Caprifoliaceae). In the discussion Willis's figures are followed by Engler's in parentheses.

As indicated earlier there is more consensus of opinion concerning genera than concerning species or families. Some of the families listed in Tables 1.1-4 are certainly not generally accepted, but the conclusions are not basically affected by the choice of system. New

Guinea has 227 (187) families in common with West Malesia, and 213 (178) with Australia, while 195 (165) families occur in all three areas. Of this latter category many are about equally represented in Malesia and Australia but there is a preponderance of families better developed in Malesia (e.g. Acanthaceae, Ericaceae,

Table 1.1 Families in West Malesia not known from New Guinea

Aceraceae	Philadelphaceae (Saxifragaceae)
Altingiaceae* (Hamamelidaceae)	Pinaceae
Ancistrocladaceae	Pistaciaceae (Anacardiaceae)
Anisophylleaceae (Rhizophoraceae)	Pyrolaceae
Antoniaceae (Loganiaceae)	Rhodoleiaceae* (Hamamelidaceae)
Berberidaceae	Ruppiaceae (Potamogetonaceae)
Betulaceae	Salicaceae
Buxaceae	Salvadoraceae
Caprifoliaceae s.s.	Saururaceae
Carlemanniaceae* (Caprifoliaceae)	Schisandraceae
Crassulaceae	Scyphostegiaceae*
Dipsacaceae s.s.	Sphenocleaceae (Campanulaceae)
Erythropalaceae* (Olacaceae)	Symphoremaceae (Verbenaceae)
Hydrophyllaceae	Taxaceae
Illiciaceae	Tetrameristaceae* (Theaceae)
Iteaceae (Saxifragaceae)	Trapaceae
Limnocharitaceae (Butomaceae)	Trichopodaceae* (Dioscoreaceae)
Lowiaceae*	Trigoniaceae
Nyssaceae	Tristichaceae (Podostemonaceae)
Pentaphyllaceae*	Valerianaceae
Petrosaviaceae* (Liliaceae)	

* Families of limited distribution, e.g. Scyphostegiacea, only in West Malesia (Borneo), and Pentaphyllaceae, West Malesia (Borneo) and South-east Asia.

Table 1.2 Families in Australia not known from New Guinea

Akaniaceae*	Gyrostemonaceae*
Anarthriaceae*	Hydrophyllaceae
Austrobaileyaceae*	Idiospermaceae* (Calycanthaceae)
Balanopaceae*	Limnocharitaceae (Butomaceae)
Baueraceae* (Saxifragaceae)	Petermanniaceae* (Dioscoreaceae)
Blepharocaryaceae* (Anacardiaceae)	Phytolaccaceae
Brunoniaceae*	Posidoniaceae (Potamogetonaceae)
Cabombaceae (Nymphaeaceae)	Ruppiaceae (Potamogetonaceae)
Cephalotaceae*	Sphenocleaceae (Caprifoliaceae)
Crassulaceae	Stylobasiaceae* (Rosaceae)
Davidsoniaceae*	Taxodiaceae
Dicrastylidaceae (Verbenaceae)	Tetracarpaeaceae* (Saxifragaceae)
Donatiaceae (Stylidiaceae)	Tetragoniaceae (Aizoaceae)
Dysphaniaceae*	Tremandraceae*
Ecdeiocoleaceae* (Restionaceae)	Tristichaceae (Podostemonaceae)
Emblingiaceae* (Capparidaceae)	Zamiaceae (Cycadaceae)
Eremosynaceae* (Saxifragaceae)	Zannichelliaceae
Eucryphiaceae	Zosteraceae (Potamogetonaceae)
Frankeniaceae	

* Families endemic to Australia, except Balanopaceae, which also occur in New Caledonia and Fiji.

Table 1.3 Families in New Guinea not known from West Malesia

Aegialitidaceae (Plumbaginaceae)	Flindersiaceae* (Rutaceae)
Atherospermataceae (Monimiaceae)	Himantandraceae*
Batidaceae	Heliconiaceae (Musaceae)
Byblidaceae*	Iridaceae
Cartonemataceae* (Commelinaceae)	Juncaginaceae
Cochlospermaceae	Philesiaceae (Liliaceae)
Corsiaceae	Plantaginaceae
Corynocarpaceae*	Sphenostemonaceae* (Aquifoliaceae)
Cupressaceae	Trimeniaceae*
Eupomatiaceae*	Xanthorrhoeaceae (Liliaceae)

* Families restricted, or nearly restricted, to New Guinea and Australia/Pacific. Most of the families of Table 1.3 comprise few species. *Aegialites* occurs in South-east Asia and the Lesser Sunda Islands and has been found fossil in Borneo (Muller 1972).

Table 1.4 Families in New Guinea not known from Australia

Averrhoaceae (Oxalidaceae)	Joinvilleaceae (Flagellariaceae)**
Balsaminaceae*	Juglandaceae
Barclayaceae (Nymphaeaceae)	Lophopyxidaceae* (Icacinaceae)
Begoniaceae*	Magnoliaceae
Bonettiaceae (Theaceae)	Marantaceae
Buddlejaceae	Mastixiaceae* (Cornaceae)
Camelliaceae (Theaceae)	Meliosmaceae (Sabiaceae)
Chloranthaceae**	Myricaceae**
Clethraceae	Pandaceae (if *Galearia* is placed in this family)
Coriariaceae**	
Crypteroniaceae	Pentaphragmataceae* (Campanulaceae)
Ctenolophonaceae (Linaceae)	
Daphniphyllaceae*	Sabiaceae*
Dipterocarpaceae*	Sarcospermataceae*
Ellisiophyllaceae (Scrophulariaceae)	Saxifragaceae s.s.
Gnetaceae*	Staphyleaceae
Heliconiaceae (Musaceae)**	Styracaceae
Hydrangeaceae (Saxifragaceae)	Triplostegiaceae (Dipsacaceae)

* Families confined to, or showing strong speciation in, Malesia.
** Occurring in New Caledonia, New Zealand, or elsewhere in the Pacific.

Gesneriaceae, Melastomataceae, Myristicaceae, Theaceae sens. lat., Urticaceae), over those better developed in Australia (e.g. Cunoniaceae, Epacridaceae, Goodeniaceae, Monimiaceae, Proteaceae, Restionaceae, Stylidiaceae).

At first sight the number of families in Tables 1.1 and 1.2 are rather well balanced but there is a notable preponderance of endemic families in Table 1.2. There is a concentration of endemic families in south-western Australia; twenty-two families do not occur north of the tropic of Capricorn. To this list might be added a few families not represented in Australia but occurring on islands of the south-west Pacific: Amborellaceae (New Caledonia), Alseuosmiaceae (Caprifoliaceae?, New Caledonia, New Zealand), Brexiaceae (Saxifragaceae, New Zealand), Degeneriaceae (Fiji), Oncothecaceae (Guttiferae?, New Caledonia), Paracryphiaceae (Eucryphiaceae?, New Caledonia), Phellinaceae (Aquifoliaceae, New

Caledonia), Strasburgeriaceae (New Caledonia). In my opinion some of the small families recently proposed for Australia and the Pacific, such as Idiospermaceae (Blake 1972) or Alseuosmiaceae, are more entitled to family rank than, for example, Pyrolaceae or Symphoremaceae in Table 1.1. Several families of Tables 1.1 and 1.2 are represented in the Moluccas, Celebes or the Lesser Sunda Islands.

The family distribution can be summarised in five points:

1. New Guinea shows no endemism at family level. The only family not represented in either West Malesia or Australia, Heliconiaceae, is almost exclusively American, only *Heliconia indica* occurring in the Old World, from the Moluccas to Samoa (Green 1969).

2. A fairly large number of Australian families, 37 (18) — or 45 (21) if the Pacific is included — does not occur in New Guinea. The number of families occurring in West Malesia but unknown from New Guinea is of comparable magnitude, 42 (24), but the floristic break between West Malesia and New Guinea is somewhat obscured by the presence in Celebes, the Moluccas or the Lesser Sunda Islands of families that are unknown from either West Malesia or New Guinea.

3. Many of the thirty-seven families in Table 1.2 are endemic, and more than half (twenty-two) are not represented in the tropics; the families in Table 1.1 are mostly widespread.

4. West Malesia and New Guinea share more families than Australia and New Guinea, most of these are either well represented in all three areas or have their strongest development in Malesia and New Guinea.

5. There are more families in common between West Malesia and New Guinea not known from Australia, than families in common between Australia and New Guinea not known from West Malesia.

These points lead to the conclusion that at family level New Guinea is more allied to West Malesia than to Australia. Nevertheless there is also a strong floristic demarcation between the floras of New Guinea and West Malesia.

Genera

For statistical purposes the genus is the most suitable unit, since there is not so much difference of opinion among taxonomists about their circumscription. The differences in published totals of New Guinea genera are due partly to the continuous finding of new records, but are also dependent on the acceptance of a genus as indigenous. I have left out of my analysis all genera represented by doubtfully native species and also those of which the occurrence in New Guinea is not sufficiently documented.

The advantage of studying the whole flora is that the result is not biased by the choice of a few spectacular cases illustrating the points one wants to demonstrate. Of course this exercise also has its drawbacks, as discussed in van Balgooy (1971). These are:

1. Our knowledge of the systematics within the various families is unbalanced. Within families treated in *Flora Malesiana* and other recent revisions the genera are generally well circumscribed, but

they are insufficiently understood in other families such as Myrsinaceae, Myrtaceae and Monimiaceae.

2. All genera are treated alike, but it is obvious that they differ in various respects: size, age, ecological tolerance, dispersal ability, etc.

3. The classification into distribution types, although established empirically, depends partly on the subjective appraisal of the author. *Nepenthes*-Nep. will be classified as Malesian by one author, since the bulk of the species occurs in Malesia, and as paleotropical by another, on account of the presence of some primitive species in the area between Madagascar and India. This problem is familiar to anyone engaged in systematic classification of organisms. In this part I have accepted five main distribution types and fifteen finer categories.

4. The use of genera obscures the distribution of infrageneric categories. The genus *Ficus*-Mor. (Corner 1965) is classified here as pantropical (type 1), but accepting as units the twelve sections represented in New Guinea these would have to be assigned to four types: eight in type 3, two in 4a, one in 7a, and one in type 8. No attempt has been made to track down the distribution of these infrageneric taxa.

The compilation of genera undoubtedly has a few inaccuracies and omissions. Very often it is difficult to establish the complete area of a genus, since many new records keep turning up, some of these hidden in the herbarium. It is, however, a matter of experience that if a flora is reasonably well known, and this is certainly true for New Guinea, later additions do not change the overall 'phytogeographic picture'.

A list was prepared of all genera represented by indigenous species in New Guinea, their complete distribution outside New Guinea was established, and notes were made regarding habit (herb, climber, tree, etc.) and habitat (mangrove, rain forest, 'alpines', etc.). Finally a spectrum was made of the main distribution types.

Figure 1.3

a —
Connarus-Conn. (after van Balgooy 1971), exemplifying a pantropical genus: type 1.

b ---
Coriaria-Coriar. (after van Steenis and van Balgooy 1966), illustrating a genus confined to the extratropical parts of both hemispheres and the tropical mountains: type 1a.

c ...
Batis-Bat. (after van Steenis and van Balgooy 1966 and Hoogland 1972), exemplifying an amphipacific tropical genus: type 2.

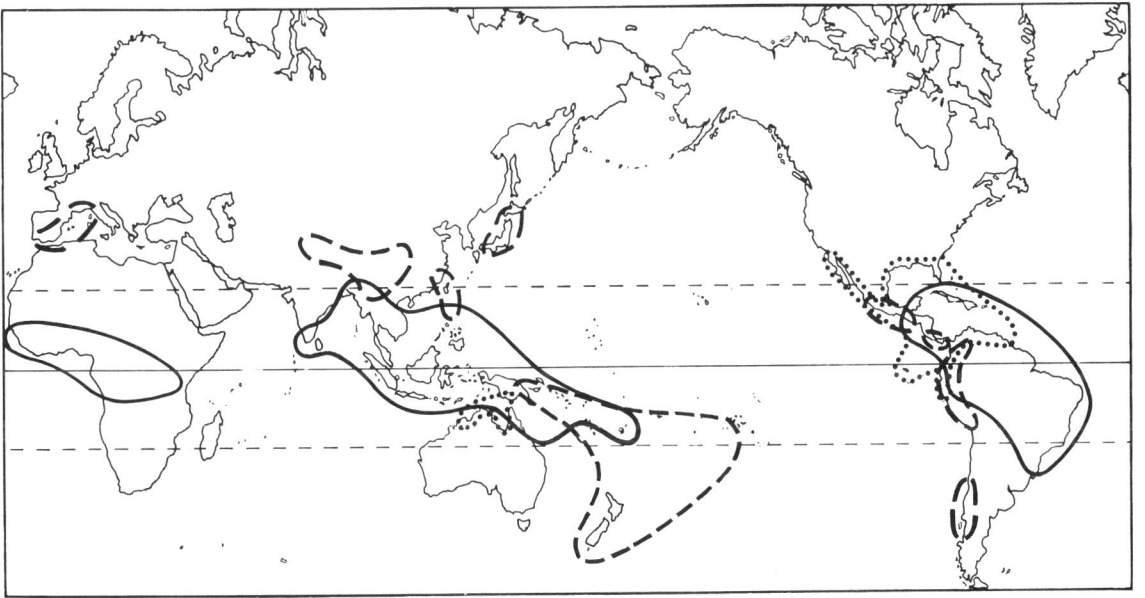

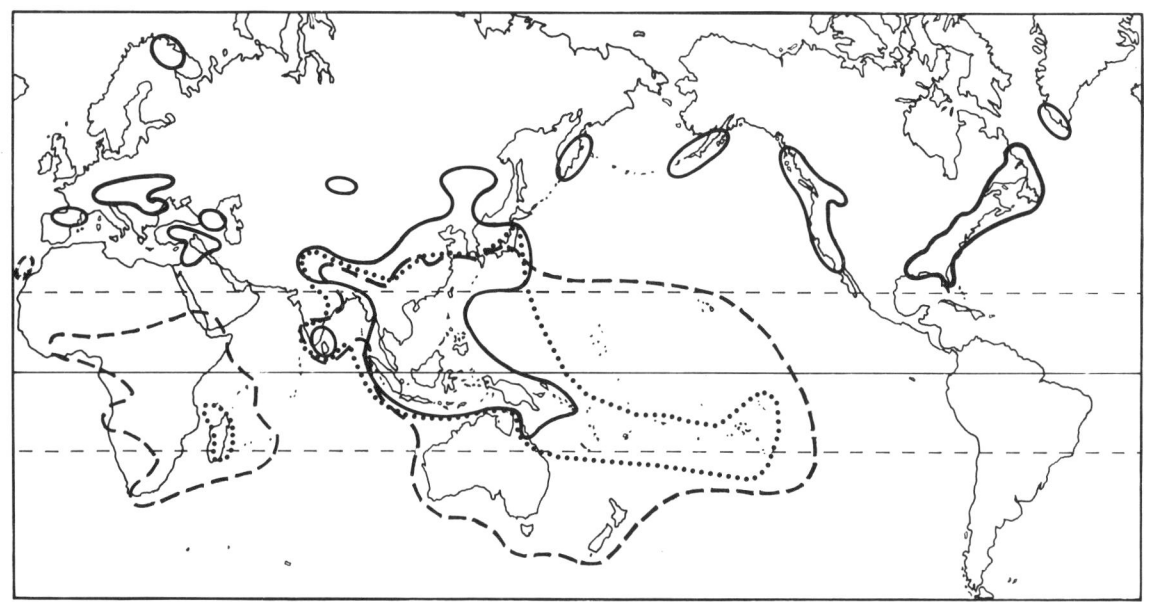

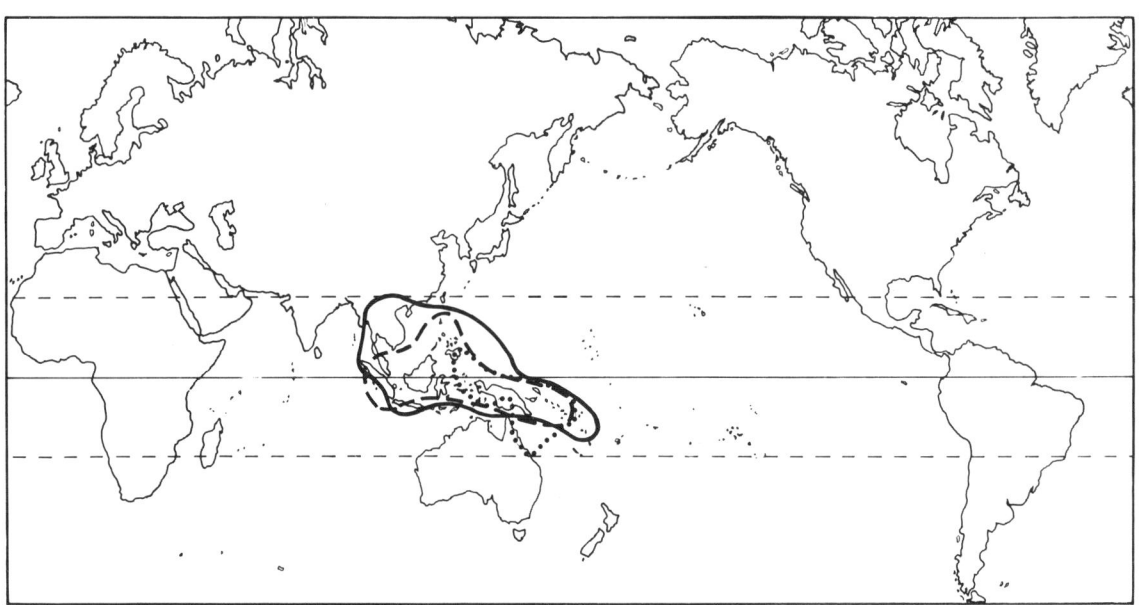

The following fifteen distribution types were established, each illustrated by a typical representative (Figs. 1.3-7).

Type 1. Cosmopolitan and pantropical genera, e.g. *Connarus*-Conn., *Cyperus*-Cyp., *Solanum*-Solan.; Fig. 1.3a.

Type 1a. Temperate wides, e.g. *Coriaria*-Coriar., *Euphrasia*-Scroph.; Fig. 1.3b.

Type 2. Amphipacific tropical genera, either centering in the Neotropics, e.g. *Heliconia*-Mus.; in the Old World, e.g. *Schismatoglottis*-Arac.; or with equal representation in both, e.g. *Batis*-Bat.; Fig. 1.3c.

Type 2a. Northern temperate, extending into the tropics, e.g. *Rhododendron*-Eric., *Euonymus*-Celast.; Fig. 1.4a.

Figure 1.4

a —
Rhododendron-Eric. (after Good 1964), a northern temperate genus, centring in east Asia and penetrating into the tropics by way of Malesia where it has a secondary centre of speciation in New Guinea (Sleumer 1966); a single species is endemic in north-east Queensland; exemplifying type 2a.

b ---
Pittosporum-Pitt. (after van Balgooy 1971), widespread in the Old World, with about equal numbers of species in the African, Asian, Australian and Pacific sectors; exemplifying type 3.

c ...
Balanophora-Balanoph. (after van Balgooy 1975), centring in East Asia and Malesia, with one species extending to Australia; exemplifying type 4.

Figure 1.5

a —
Kopsia-Apoc. (after van Steenis and van Balgooy 1966), as *Balanophora*, centring in South-east Asia and West Malesia; one species extends to New Guinea and Melanesia but none to Australia; exemplifying type 4a.

b ---
Dendromyza-Sant. (after van Steenis and van Balgooy 1966), centring in Malesia, reaches the Solomons in the east but is not represented in either Asia or Australia; illustrating the distribution of a Malesian genus: type 5.

c ...
Archidendron-Leg. (after van Balgooy 1975), although not confined to New Guinea, has a clear centre of speciation there; some species found in adjacent islands and in Queensland; exemplifying type 6.

Type 3. Old World generally, not centering in either Africa, Asia, Malesia or Australia, e.g. *Pittosporum*-Pitt. Paleotropical genera not present in Africa are included here also, e.g. *Aegiceras*-Myrs.; Fig. 1.4b.

Type 4. Paleotropical and Indo-Malesian genera, distinctly centering in Africa, Asia or Malesia, and reaching Australia with a single or very few species, e.g. *Balanophora*-Balanoph., *Melastoma*-Melast., *Nepenthes*-Nep.; Fig. 1.4c.

Type 4a. As above, but not represented in Australia, e.g. *Kopsia*-Apoc., *Shorea*-Dipt.; Fig. 1.5a.

Type 5. Strictly Malesian genera, e.g. *Dendromyza*-Sant., also those clearly centering in Malesia but reaching Asia and/or Australia with a single species, e.g. *Cyrtandra*-Gesn.; Fig. 1.5b.

Type 6. Subendemic genera, centering in New Guinea but having some species also in adjacent areas: Moluccas, Solomons, Queensland, etc., e.g. *Archidendron*-Fab., *Corsia*-Cors.; Fig. 1.5c.

Type 6a. Strictly endemic genera, e.g. *Annesijoa*-Euph., *Sericolea*-Elaeoc.; Fig. 1.6a.

Type 7. Genera centering in Australia but also represented elsewhere, e.g. *Hibbertia*-Dill., *Myoporum*-Myop.; Fig. 1.6b.

Type 7a. Genera (virtually) confined to Australia and New Guinea, e.g. *Eupomatia*-Eup., *Stackhousia*-Stack.; Fig. 1.6c.

Type 8. Genera centering in the Pacific, e.g. *Ascarina*-Chlor., *Bikkia*-Rub.; Fig 1.7a.

Type 8a. Genera equally well represented in Australia and the Pacific, e.g. *Quintinia*-Sax./Esc.; Fig. 1.7b.

Type 9. Subantarctic/southern hemisphere genera, e.g. *Astelia*-Lil., *Nertera*-Rub., *Nothofagus*-Fag.; Fig. 1.7c.

The distribution type spectrum of New Guinea genera based on this classification is as shown in Table 1.5. The types can be grouped into five supra-categories (van Steenis 1954; Keng 1970) as shown in Table 1.6. Table 1.6 shows that there are more than twice the

Table 1.5 Number of genera with indigenous species in New Guinea per distribution type

Distribution type	Number of genera	Percentage
1	307	21.0
1a	39	2.7
2	30	2.0
2a	27	1.8
3	294	20.1
4	89	6.1
4a	221	15.1
5	82	5.6
6	97	6.6
6a	98	6.7
7	36	2.5
7a	47	3.2
8	24	1.6
8a	46	3.1
9	28	1.9
Total	1465	100.0

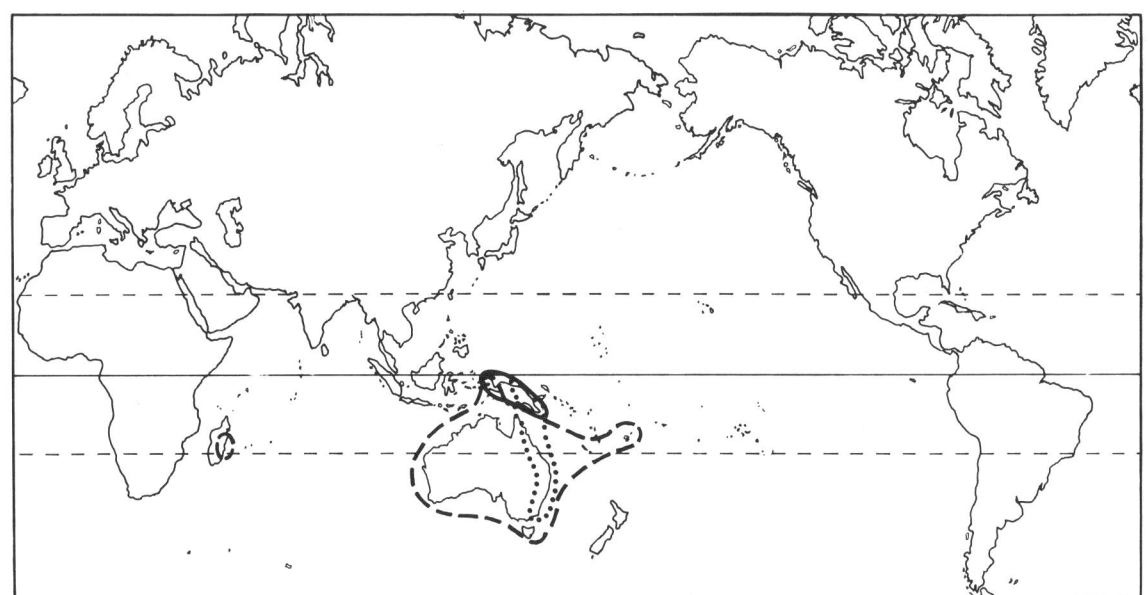

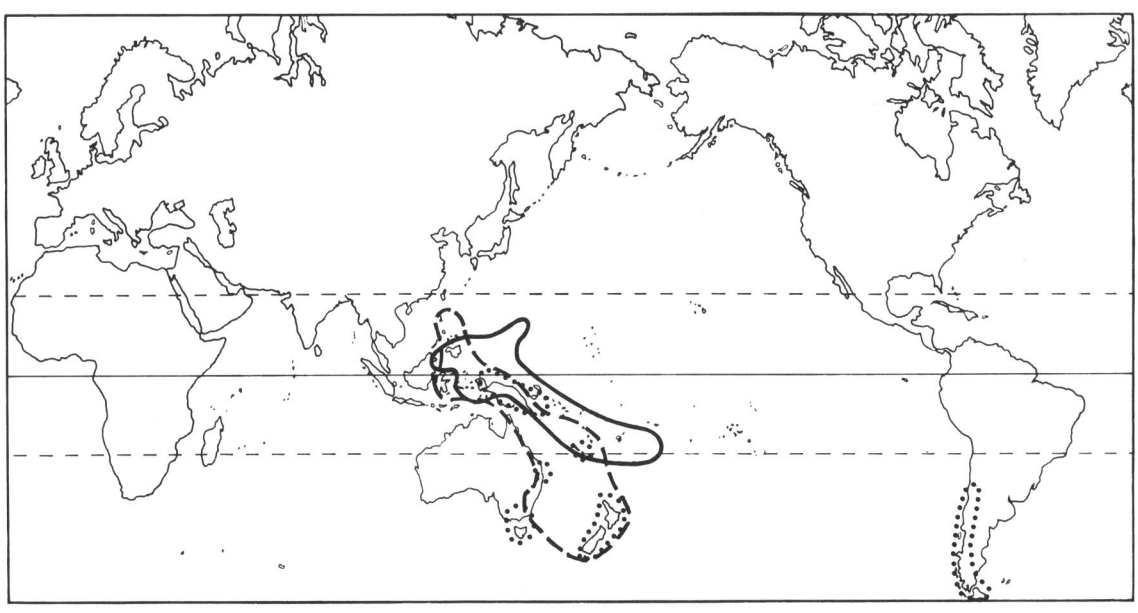

number of genera centering in Malesia, Asia and other countries to the north and west of New Guinea, than those centering in Australia, the Pacific and the southern hemisphere generally. This is in agreement with all other previous authors who have analysed the Phanerogam flora of New Guinea, even though their criteria of classification may have been different (Lam, van Steenis, Good, Hoogland). On the whole the flora of New Guinea is distinctly more allied to that of Indo-Malesia than to that of Australasia. This is further supported by a few other facts. A great number of genera, type 4a, most of type 5 and several of other types, terminate their eastward distribution rather abruptly in New Guinea, or, as in the case of *Rhododendron,* have produced a great number of species in New Guinea and are practically absent in Australia (one species in

Figure 1.6

a —
Sericolea-Elaeoc., confined (endemic) to New Guinea; illustrating type 6a.

b ---
Hibbertia-Dill. (after Hoogland 1951, 1972), a genus with a centre of speciation in Australia; a fairly high number of species outside the continent, mostly New Caledonia; one species has been found in Madagascar, and two species cross the Torres Strait; exemplifying type 7.

c ...
Eupomatia-Eupom. (after Hoogland 1972), a genus of two species, one confined to Australia (Queensland) and the second extending to New Guinea; exemplifying type 7a.

Figure 1.7

a —
Bikkia-Rub. (after van Steenis and van Balgooy 1966), extending from north-east Borneo to Niue Island in the Pacific, centring in the Pacific (New Caledonia); exemplifying type 8.

b ---
Quintinia-Saxif./Escal. (after van Balgooy 1971), represented equally strongly in the Pacific and in Australia and extending north to East Malesia; exemplifying type 8a.

c ...
Nothofagus-Fag. (after van Steenis and van Balgooy 1966), a southern hemisphere genus extending into the tropics in the south-west Pacific (New Caledonia and New Guinea); illustrating type 9.

Table 1.6 Number of genera per distribution category

Category	Number of genera	Percentage
A. Widespread genera (types 1, 1a and 3)	640	43.7
B. Tropical amphipacific genera (type 2)	30	2.0
C. Genera centering in countries to the north and west of New Guinea (types 2a, 4, 4a and 5)	419	28.6
D. Genera confined (endemic) to, or centering in, New Guinea (types 6 and 6a)	195	13.3
E. Genera centering in countries south or east of New Guinea (types 7, 7a, 8, 8a and 9)	181	12.4
Total	1465	100.0

Rhododendron). On the other hand, many southern and eastern genera do not have New Guinea as their endstation, but are found further west in the Moluccas, Celebes, Borneo, Philippines or even South-east Asia (*Ascarina, Nertera, Oreobolus, Phyllocladus*).

New Guinea shares *circa* 1120 genera with West Malesia and *circa* 900 with Australia, which means that 345 New Guinea genera do not occur in West Malesia against 565 New Guinea genera that are unknown from Australia. Van Steenis (1954) who accredited New Guinea with 1094 genera found that 563 of these did not cross the Torres Strait, whereas 340 tropical Australian genera were unknown from New Guinea. Hoogland (1972), accepting 1501 native genera for New Guinea, states that 608 of these are absent from Australia; of 1604 Australian genera 711 are said to be absent from New Guinea. This last figure, however, includes genera confined to the extratropical parts of Australia; of the tropical Australian genera only 205 are absent from New Guinea. Although it is quite clear that there is a considerable floristic break at Torres Strait, it must be pointed out that there is also a floristic discontinuity between the floras of New Guinea and of West Malesia, a discontinuity of the same magnitude as those between the floras of South-east Asia and West Malesia and between Taiwan and the Philippines.

Van Steenis (1950) divided Malesia, as defined earlier, into three parts: West Malesia (Greater Sunda Islands and Philippines), South Malesia (Java and Lesser Sunda Islands) and East Malesia (New Guinea, Celebes and Moluccas). It was later shown (van Balgooy 1960, 1971) that almost the whole tropical Pacific floristically belongs to East Malesia. The most logical division of the Indo-Malesian region would be to accept three subregions: South-east Asia, West Malesia comprising van Steenis's West and South Malesia (see Fig. 1.1), and the Papuan subregion comprising New Guinea, Celebes, Moluccas and the tropical Pacific and excluding New Caledonia.

The broad outline of floristic demarcation is clear, but closer study of the distribution patterns reveals some interesting facets worth mentioning, since they are obscured in the overall picture. There are thirty genera belonging to type 2, the amphipacific tropical genera. Although small in number, their presence on both sides of the Pacific poses a major problem in phytogeography, one fully discussed by van Steenis (1962a). It is remarkable that some of the endemic or subendemic genera show neotropical relationships: *Corsia*-Cors., *Annesijoa*-Euph., *Eriandra*-Polygal.

Although New Guinea has no endemic families there is an impressive number of endemic genera. Their number was formerly believed to be even higher, but several genera have in the meantime been reduced to synonyms of genera with a wider distribution, or have been found outside New Guinea. Many of the endemic genera show vigorous speciation and they belong to a wide range of families, suggesting that evolution has been going on for a considerable time. Widespread genera of northern and western derivation (types 2a, 4, 4a, and 5) are better represented in number of species than those of southern and eastern origin (types 7, 7a, 8, 8a, and 9). Ten genera of type 8 (Pacific genera) do not occur in New Guinea proper (e.g. *Crossostylis, Joinvillea*) but do occur in the Solomons.

One would expect that Australia-based genera (types 7 and 7a) would be better represented in the parts of New Guinea that have a seasonal climate, i.e. the extreme south and the Port Moresby area, but Heyligers (1972c) has pointed out that the generic distribution spectrum for Port Moresby is hardly different from that of New Guinea as a whole.

Webb and Tracey (1972) compared rain forest sites on either side of Torres Strait and found that in general they show a fairly large percentage of shared genera. However, without comparisons with similar sites on the Sunda Islands we cannot say whether the New Guinea rain forest is more allied to that of Queensland or to that of West Malesia. From my data it appears that out of about 750 New Guinea lowland and lower montane rain forest genera about 400 are unknown from north-east Queensland. Conversely, from a list of 430 genera of the north-east Queensland rain forest, which the present writer compiled with the help of Hyland and others, there are about ninety unknown from New Guinea.

Van Steenis has remarked that the flora of the lowlands may tell a different tale from that of the mountains. For the Lesser Sunda Islands Kalkman (1955) found that both the lowland and the montane flora were derived mainly from the west.

The genera found exclusively, or nearly so, in montane habitats in New Guinea are listed in Table 1.9 at the end of this part. Only exceptionally do they come down below the 1000 m limit which is considered by van Steenis (1934-6) as the boundary between the tropical lowland and montane zones. Genera that are best developed in the mountains but have a few lowland species, or species descending to the lowlands, for example *Perrottetia*-Cel., *Rhododendron*-Eric., *Weinmannia*-Cunon., are not included in Table 1.9. The genus *Citriobatus*-Pitt., which has been recorded at 1680 m in New Guinea (Schodde 1972), but occurs in the lowlands elsewhere in Malesia, has also been omitted. The distribution type spectra of the montane genera listed in Table 1.9 are given in Tables

1.7 and 1.8. Bracketed figures refer to totals when the less strictly montane genera (*Perrottetia*, etc.) are included.

Table 1.7 Number of montane genera per distribution type

Distribution type	Number of genera	Percentage
1	3 (14)	2.2 (5.8)
1a	32 (36)	23.2 (14.9)
2	1 (5)	0.7 (2.1)
2a	15 (24)	10.9 (9.9)
3	4 (16)	2.9 (6.6)
4	2 (4)	1.5 (1.7)
4a	11 (26)	8.0 (10.7)
5	— (3)	— (1.2)
6	12 (24)	8.7 (9.9)
6a	16 (28)	11.6 (11.6)
7	6 (9)	4.3 (3.7)
7a	4 (7)	2.9 (2.9)
8	6 (7)	4.3 (2.9)
8a	9 (16)	6.5 (6.6)
9	17 (23)	12.3 (9.5)
Total	138 (242)	100.0 (100.0)

Table 1.8 Number of montane genera per distribution category

Category (see Table 1.6)	Number of genera	Percentage
A	39 (66)	28.3 (27.3)
B	1 (5)	0.7 (2.1)
C	28 (57)	20.3 (23.5)
D	28 (52)	20.3 (21.5)
E	42 (62)	30.4 (25.6)
Total	138 (242)	100.0 (100.0)

The southern affinities of the montane flora of New Guinea have long been recognised (e.g. by Diels 1930) but have never been expressed in figures. Lam (1934) stated that the Asiatic element in the flora of New Guinea prevailed even in the mountains, whereas according to Schodde and Calaby (1972) the mountain flora is dominated by southern taxa. According to the present analysis there is indeed a preponderance of southern and eastern genera over northern and western-derived ones, but it is only very slight if montane genera are taken in a wide sense. If the eurytherm genera (i.e. those occurring in the mountains as well as in the lowlands) had been included in this analysis, the western/northern genera would outnumber the eastern/southern genera. The inclusion of New Guinea in the Indo-Malesian region is therefore determined by the preponderance of western-derived taxa which dominate the lowlands.

It is interesting to note that strictly Malesian and strictly Australian distribution types are very poorly represented in Table 1.7. Best represented are genera with a wide distribution in both hemispheres (type 1a), northern hemisphere genera (type 2a), Indo-

Malesian genera (type 4a), subendemic genera (type 6a) and southern hemisphere genera (type 9).

Species

The state of our knowledge of the New Guinea flora is not such that an analysis along the same lines as for the families and genera is justified. No attempt has therefore been made in this part to investigate species distribution patterns. The following data have been taken mainly from the literature.

Good (1960) has estimated the Angiosperm flora of New Guinea to comprise 9000 species of which he considered about 90 per cent endemic. Thus the species is a less suitable unit to establish outward relations of the New Guinea flora. Floristic divisions inside New Guinea should await a better understanding of the delimitation and distribution of the species. Also details of the past (recent) history and ecological studies should be based on the species.

Hoogland (1972) remarks that although Torres Strait forms a strong barrier to many families and genera there are several species occurring on both sides of the Strait. As expected these are best represented in the coastal habitats including mangrove swamps. Ecological conditions in mangroves on both sides of the strait are very similar, most species are adapted to dispersal by sea water and have a wide distribution. A relatively small percentage of species of savanna woodland and monsoon forest, a type of habitat best developed in Australia, is represented in New Guinea where this habitat type is found only in the south and south-east and in a few isolated spots elsewhere. Very few New Guinea species of seasonal habitats are absent from Australia. Conversely, there are very few New Guinea lowland rain forest species that cross the Strait to Australia, where rain forest occurs in small scattered areas. There are, however, a number of montane species of New Guinea that are also known from Queensland and from localities at higher latitudes in south-eastern Australia, Tasmania and New Zealand.

Heyligers (1972c), analysing the flora of the Port Moresby area, found that of a total of 292 species about one-third did not occur elsewhere in New Guinea, which points to a long history of seasonal climate in this area. Most of the widespread species came via dry areas in Asia/Malesia and Australia before the Pliocene mountain-building period. New migrations took place during Pleistocene glacial periods which favoured mainly Australian and Malesian drought-loving species. A rise of sea level and increased rainfall in the Torres Strait area after the last glacial eliminated many monsoonal species in southern New Guinea.

Origin of the New Guinea Flora

So far the floristic facts on the flora of New Guinea have been presented without much comment on its origin. Speculations on the past history of a flora or fauna on the basis of present-day distribution is obviously a hazardous pastime, yet it forms the *piece de résistance* in any biogeographical paper.

Lately there has been a spate of papers on plate tectonics and continental drift, see for example Smith, Briden and Drewry (1973) and the numerous references in Raven and Axelrod (1972) and in Whitmore (1973). Several authors, including Raven and Axelrod, have attempted to explain biogeographic relationships in the south-west Pacific in the light of reconstructions of land and sea by geophysicists. In a symposium on the role of Torres Strait as a bridge and barrier to migration (Walker 1972) ample attention was paid to plate tectonics and attendant phenomena.

In a nutshell the supposed series of events in this part of the world are as follows. In the Middle Cretaceous (*circa* 100 million years B.P.) Australia formed part of a huge southern landmass, Gondwanaland, which further comprised Africa, Madagascar, part of India, Antarctica and South America. The eastern margin of this supercontinent started to break off in the Upper Cretaceous some 80 million years ago. This huge fragment comprising Australia and land masses now partly or wholly submerged (New Zealand, New Caledonia, Lord Howe Rise and Queensland Plateau) moved north and eastwards. Contacts with South America persisted until the Eocene 45-40 million years B.P. The north-east moving Australian Plate was met by the west moving Pacific Plate, and this caused extensive faulting and the formation of island arcs, which spread eastwards in front of Australia. Emergence of lowland areas of New Guinea took place at about 10° S during the Miocene (25 million years B.P.). The high mountains of New Guinea were not formed until the Pliocene as a result of the collision between the Australian and Asian Plates, which also caused the foundering of the Queensland Plate. Anti-clockwise rotation of New Guinea separated it from Queenland and gave the island its present shape and position.

Walker, and especially Raven and Axelrod, find the biogeographic facts to fit this reconstruction exactly. In Walker's words (1972: 401): 'The broad outlines of the distributions of most plant and animal groups between New Guinea and Australia are consistent with what is now thought to have happened to the distributions of land and sea in the geological history of the southern hemisphere.'

It is supposed that the original temperate Gondwana flora was gradually decimated upon entering the tropical zone. Only plants that could adapt themselves to the increasing aridity and those that found refuge in the south-east corner of Australia and in the mountains of New Guinea survived. Invasion of Malesian rain forest elements is thought to be a recent pheonomenon.

> ... at the beginning of the Quaternary, the Cape York Peninsula and its northward extension which is now southern Papua had available to it: parts, at least, of an old Gondwanic biota; a perhaps evolutionarily younger but already well-established Australian biota of mixed origins; newly immigrating groups from Asia. This was a setting in which major taxa which had never before met came face-to-face and leaf-to-leaf (Walker 1972: 403).

These reconstructions of the geological events, though they are attractive at first sight, and though they explain a number of facts, leave certain details unexplained. If New Guinea is the scene of recent floristic mixing of newly arrived Asiatic/Malesian immigrants and remnants of a Gondwana flora, then the high endemism of the flora at genus and species levels is hard to explain. The more so since this endemism is not confined to taxa of southern affinity as would be expected, but is also found in taxa of Asiatic affinity. Furthermore, if the strong speciation in New Guinea is a recent response to the presentation of empty niches it is strange that the same did not happen in, for example, the Solomons, where Whitmore (1969: 563) found reasons to state: 'So far we can say that the evidence from the Solomon Islands is that, in the humid tropics, flowering plant evolution has been very slow, even in a region undergoing vigorous, one might almost say dramatic, geological change.'

The strong discontinuity of the floras of New Guinea and Australia, although far less dramatic if ecologically comparable parts of the flora are contrasted instead of regional ones, remains a reality. There is also a strong discrepancy with West Malesia; this has perhaps been insufficiently stressed in the past. It shows that the New Guinea flora has undergone independent development, and judging from the high percentage of endemism and the fact that many families are better represented in New Guinea than anywhere else, this cannot be very recent.

Smith (1970) has pointed out that there is a concentration of primitive families in the area between India-Japan and Fiji-New Zealand, with many families like Annonaceae, Lauraceae, Nymphaeaceae and others in common between Asia/Malesia and the south-west Pacific. Early authors, Hooker (1860) and Diels (1934, 1936), have already pointed to the presence in the south-west Pacific of many endemic taxa of apparently northern and tropical derivation: for example Rutaceae-Boronieae, Casuarinaceae (to which might be added *Blepharocarya* in Australia, Myricaceae and taxa of Theaceous stock in New Caledonia). *Nothofagus,* the only southern hemisphere genus of the family Fagaceae, a genus with a well documented fossil record and proven poor dispersal capacity and limited ecological tolerance (van Steenis 1971), was already present in Cretaceous deposits in Australia and New Zealand, and there is no indication that it came there by way of Africa or South America, instead it appears to be of Asiatic derivation. The Proteaceae are most likely of tropical origin (Johnson and Briggs 1963).

It has now been convincingly shown that the high montane flora of New Guinea is more allied to that of the southern countries than to that of Asia, a situation quite contrary to that of the lowland flora. It should be noted that most of the southern elements in the New Guinea montane flora do not belong to typically Australian families but to families of wide southern distribution, for example Cunoniaceae and Monimiaceae.

The relationship between Asia/Malesia and the south-west Pacific is apparently not modern as a whole, although part of it is. The savanna element in New Guinea is doubtlessly young and montane taxa may have taken the opportunity to expand their area when, during Pleistocene glaciation, suitable habitats were more extensive than now. Apparently distances to New Guinea were shorter for the southern elements than for the Asiatic ones. Thus the floristic composition of the south-west Pacific generally suggests that there is a chequered layer of old north-south connections underlying modern migrations.

The distinctiveness of the Australian flora, its distant affinity with South Africa and its stronger affinity with South America is consistent with the land reconstructions but, as Whitmore (1973) pointed out, two more problems—not directly bearing on New Guinea—are not satisfactorily explained: the transpacific tropical relationships and the high endemism of archaic forms of both northern and southern derivation in New Caledonia. Current reconstructions of continental movements have in common that the position of Antarctica is considered a fixed point. Since all movements of the other continents are relative to this fixed point they may be exaggerated. Australia may have reached tropical latitudes much earlier than is generally believed (Smith 1970).

The role of the land masses to the north and east of the Australian Plate is unknown. If early contacts between Asia/Malesia and the south-west Pacific existed—and I think the evidence for this is strong—it was probably not with Australia itself but with its bordering land masses: Lord Howe Rise, New Caledonia and the Queensland Plateau.

Most botanists occupying themselves with plant distributions in the south-west Pacific have on various grounds made claims for more land or land connections to explain the facts of distribution. They include Corner (1963, *Ficus*), Smith (1945, Winteraceae; 1970, primitive Angiosperm families), van Steenis (1962a, transpacific affinities), Whitmore (1969, Solomons flora) and van Balgooy (1971, Pacific Phanerogam genera). These claims are supported by some form of continental displacement in that chance dispersal as an all-explaining solution can be denied, but some aspects of plant distribution in relation to the distribution of land and sea in the south-west Pacific, including New Guinea, are, as yet, not clear.

Table 1.9 Annotated list of mountain genera

Family	Genus	Distribution type	Notes
Angiosperms			
Apiaceae	*Oreomyrrhis*	9	Alp. herb
(Umbelliferae)	*Trachymene*	7	Herb, alp. and forest
Araliaceae	*Palmervandenbroeckia*	6a	Tree, allied to Paleotropic genera
Asteraceae	*Abrotanella*	9	Alp. herb
(Compositae)	*Anaphalis*	2a	Alp. and subalp. shrub and herb
	Arrhenechthites	6	Shrub, forest; one sp. in N.S.W.
	Brachyonostylum	6a	Alp. herb; allies in Pacific region
	Brachycome	8a	Subalp. herb
	Conyza	1	Herb, open places
	Cotula	1a	Alp. herb
	Erechtites	9	Herb, alp. and forest
	Ethulia	1	Herb, open places
	Ischnea	4a	Herb, alp. bogs
	Keysseria	6	Alp. herb
	Lactuca	4a	Alp. herb
	Myriactis	4a	Alp. and subalp. herb
	Olearia	8a	Shrub, alp. and forest
	Piora	6a	Alp. shrub; allied to *Tetramolopium*

Family	Genus	Distribution type	Notes
	Senecio (incl. *Bedfordia*)	1a	Herb and shrub, alp. and forest
	Tetramolopium	8	Alp. herb and shrub
Boraginaceae	*Crucicaryum* (doubtful genus)	6a	Alp. herb; allied to widespread genera
	Myosotis	1a	Alp. and subalp. herb
	Trigonotis	4a	Alp. herb
Brassicaceae (Cruciferae)	*Cardamine*	1a	Alp. and subalp. herb
	Papuzilla	6a	Alp. shrub
Callitrichaceae	*Callitriche*	1a	Water plant
Campanulaceae	*Peracarpa*	4a	Alp. herb
Caprifoliaceae	*Sambucus*	2a	Shrub, forest
Caryophyllaceae	*Cerastium*	2a	Alp. herb
	Sagina	2a	Alp. herb
	Scleranthus	1a	Alp. herb
	Stellaria	1a	Alp. herb
Centrolepidaceae	*Centrolepis*	7	Herb, alp. bogs
	Gaimardia	9	Herb, alp. bogs.
Chloranthaceae	*Ascarina*	8	Tree, forest
Coriariaceae	*Coriaria*	1a	Shrub, forest
Cunoniaceae	*Acsmithia*	8	Tree, forest; centres in New Caledonia
	Pullea	8a	Tree, forest
	Spiraeopsis	6	Tree, forest
Cyperaceae	*Carpha*	9	Sedge, alp. bogs
	Oreobolus	9	Sedge, alp. bogs
	Uncinia	9	Subalp. sedge
Daphniphyllaceae	*Daphniphyllum*	4a	Tree and shrub, forest
Dipsacaceae	*Triplostegia*	4a	Alp. herb
Elaeocarpaceae	*Sericolea*	6a	Alp. and subalp. tree and shrub
Epacridaceae	*Decatoca*	6a	Subalp. shrub; Aust. affinity
	Trochocarpa	7	Alp. and subalp. shrub
Ericaceae	*Agapetes*	4	Climb. and shrub, alp. and forest
	Dimorphanthera	6	Climb. and shrub, alp. and forest
	Gaultheria	1a	Climb. and shrub, alp. and forest
Fagaceae	*Nothofagus*	9	Tree, forest
Gentianaceae	*Gentiana*	1a	Alp. herb; N.G. sp. of Asiatic alliance
	Swertia	2a	Herb, open places
Geraniaceae	*Geranium*	1a	Alp. and subalp. herb
Gesneriaceae	*Oxychlamys*	6a	Climber, forest; allied to Indo-Malesian *Aeschynanthus*
Haloragaceae	*Gunnera*	9	Herb, forest
Icacinaceae	*Hartleya*	6a	Tree, forest; allied to *Gastrolepis* of New Caledonia
Iridaceae	*Libertia*	9	Herb, forest
	Patersonia	7a	Herb, forest
Juncaceae	*Juncus*	1a	Herb, wet open places
	Luzula	1a	Herb, wet open places
Leguminosae	*Papilionopsis*	6a	Herb, open places; Aust. affinity?
Liliaceae	*Astelia*	9	Herb, alp. bogs; N.G. sp. of Aust. affinity
Loranthaceae	*Korthalsella*	3	Parasitic shrub
Monimiaceae	*Dryadodaphne*	6	Tree, forest; Aust. affinity
	Palmeria	7a	Climber, forest
Myrtaceae	*Mearnsia*	8	Tree, forest
	Xanthomyrtus	6	Tree and shrub, alp. and forest
Onagraceae	*Epilobium*	1a	Alp. herb

Family	Genus	Distribution type	Notes
Orchidaceae	*Acianthus*	8a	Herb, open places
	Dryadorchis	6a	Epiphyte, forest
	Epiblastus	6	Epiphyte, forest
	Kerigomnia	6a	Epiphyte, forest
	Microtatorchis	6	Epiphyte, forest
	Pedilochilus	6	Epiphyte, forest
	Pterostylis	7	Herb, alp. and forest
	Sepalosiphon	6a	Epiphyte, forest
	Spiranthes	1	Herb, forest
	Thelymitra	7	Alp. herb
Oxalidaceae	*Oxalis*	1a	Herb, forest; N.G. sp. of sub-antarctic distribution
Plantaginaceae	*Plantago*	1a	Herb, alp. bogs
Poaceae	*Agrostis*	1a	Alp. grass
	Anthoxanthum	2a	Alp. grass
	Brachypodium	2a	Alp. and subalp. grass
	Bromus	1a	Alp. grass
	Coelachne	3	Grass, open places
	Danthonia	1a	Alp. grass
	Deschampsia	1a	Alp. grass
	Deyeuxia	2a	Alp. grass
	Dichelachne	8a	Alp. grass
	Echinopogon	7	Grass, open places
	Festuca	1a	Grass, alp. and forest
	Hierochloë	1a	Alp. grass
	Microlaena	3	Grass, alp. and forest
	Monostachya	6	Grass, alp. bogs
	Muehlenbergia	2a	Grass, open places
	Poa	1a	Alp. grass
	Tripogon	4	Grass, open places
	Trisetum	2a	Alp. grass
Portulacaceae	*Montia*	1a	Water plant
Primulaceae	*Lysimachia*	1a	Herb, forest
	Primula	2a	Herb, open places
Rafflesiaceae	*Langsdorffia*	9	Root parasite, forest
	Mitrastemon	2	Root parasite, forest
Ranunculaceae	*Ranunculus*	1a	Alp. herb
	Thalictrum	2a	Herb, open places
Rosaceae	*Acaena*	9	Alp. and subalp. herb
	Potentilla	1a	Alp. and subalp. herb
Rubiaceae	*Amaracarpus*	6	Shrub, forest
	Coprosma	8a	Alp. and subalp. herb and shrub
	Galium	1a	Herb, forest
	Neanotus	3	Herb, open places
	Nertera	9	Herb, forest and alp. bogs
Rutaceae	*Evodiella*	7a	Tree, forest
Saxifragaceae	*Astilbe*	2a	Herb, open places
	Carpodetus	8	Shrub and tree, forest
	Quintinia	8a	Shrub and tree, forest and alp.
Scrophulariaceae	*Detzneria*	6a	Alp. shrub; allied to *Parahebe*
	Ellisiophyllum	4a	Herb, forest
	Euphrasia	1a	Alp. herb
	Parahebe	9	Herb and shrub, alp. and forest
	Veronica	2a	Herb, open places and forest
Sparganiaceae	*Sparganium*	2a	Herb, open places and forest
Sphenostemonaceae (Aquifoliaceae)	*Sphenostemon*	8a	Tree, forest

Family		Distribution type	Notes
Theaceae	*Archboldiodendron*	6a	Tree; allied to Indo-Malesian *Eurya*
Thymelacaceae	*Drapetes*	9	Shrub, alp. bogs
Trimeniaceae	*Piptocalyx*	7a	Climber, forest
	Trimenia	8	Tree, forest
Urticaceae	*Chamaibainia*	4a	Herb, forest
	Gibbsia	6a	Herb, forest; allied to Indo-Malesian *Leucosyke*
	Lecanthus	4a	Herb, forest
	Parietaria	1a	Herb, forest
	Urtica	1a	Herb, forest
Violaceae	*Viola*	1a	Herb, forest and alp.
Winteraceae	*Drimys*	9	Tree and shrub, alp. and forest
Gymnosperms			
Cupressaceae	*Papuacedrus*	6	Tree, allied to *Libocedrus* of New Caledonia and N.Z.
Podocarpaceae	*Dacrycarpus*	4a	Tree, forest
	Phyllocladus	8a	Tree, forest

PART II

Vegetation

K. Paijmans

Introduction

This account of the vegetation of Papua New Guinea is based mainly on field observations carried out during land resources surveys over the last twenty years by plant ecologists of the Division of Land Use Research, CSIRO, Canberra. The results of these surveys have been published in a number of reports (Heyligers 1965, 1967, 1972a, 1972b; Paijmans 1967, 1969, 1971, 1973b; Robbins and Pullen 1965; Robbins 1959, 1968, 1970; Saunders 1957; Taylor 1964a, 1964b). As our knowledge of the vegetation is still somewhat sketchy, the present account is inevitably incomplete. However, the main vegetation types and their habitats can be described by matching and piecing together the survey information.

Figure 2.1
Topographic map of New Guinea

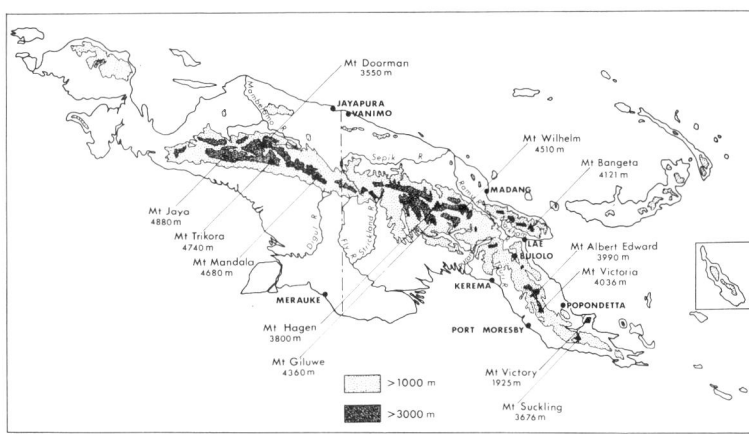

My views on the New Guinea environment have taken shape through numerous discussions and arguments with colleagues of different disciplines. I wish to thank in particular Dr D. H. Blake, formerly of CSIRO, for his thorough and sound comments and criticisms of the draft. I am grateful also to Professor D. Walker of The Australian National University, to Mr H. A. Haantjens and fellow ecologist Dr P. C. Heyligers of the Division of Land Use Research, to Mr R. Pullen of the Herbarium Australiense, and to Mr K. J. White, Assistant Director of the Department of Forests, Port Moresby, for their comments and helpful suggestions. Special thanks are due to Papuan Forest Assistant and tree namer Cappock Mago, who in his unassuming way has given me invaluable all-round assistance during field work, and on more than one occasion has saved me from becoming lost in the bush.

The great diversity of drainage conditions, climate, and altitude has led to a first grouping of the plant communities according to major environments. A second division is based on structural form, and botanical names are used to distinguish communities that share the same major environment and have the same structural form, but have different dominant species. Both in the classification and throughout the text environmental, structural and floristic characteristics are intentionally linked together. The use of botanical names is restricted, and generally the only species mentioned are those that are the most common or characteristic of each vegetation type. More comprehensive floristic lists are given in the regional survey reports.

This part deals with the terrestrial vegetation of mainland Papua New Guinea and the larger islands, and to a lesser extent with the relatively little known flora of the numerous small islands. A section on Irian Jaya concludes the part. Information on the smaller islands is taken from Brass (1956, 1959), and from Gillison (pers. comm.). The history of botanical exploration and collections in New Guinea is described in *Flora Malesiana* (de Wit 1949; van Steenis-Kruseman 1950, 1958). The geographical distribution of the structural forms described in this account is shown on the vegetation map of Papua New Guinea, scale 1:1,000,000, published by CSIRO, Division of Land Use Research (Paijmans 1975), which is available on request.

Major Environments and Structural Forms

The main environments in Papua New Guinea (Fig. 2.2) and their approximate percentages of cover are:

Coastal beach ridges and flats	½
Coastal saline and brackish swamps	1½
Lowland fresh water swamps, mainly 0-50 m a.s.l.	11
Lowland alluvial plains and fans, mainly 0-500 m a.s.l.	15
Foothills and mountains below 1000 m a.s.l.	43
Lower montane zone, 1000-3000 m a.s.l.	25
Upper montane zone, 3000-4000+ m a.s.l.	4

Figure 2.2
Major environments of Papua New Guinea

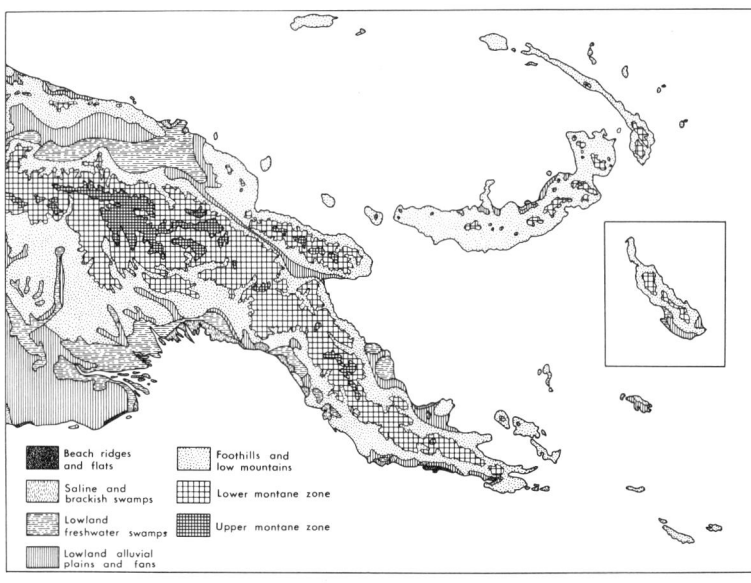

In the lowland swamps and plains drainage conditions, water regime and type of water are the main factors controlling the vegetation. Along the coast a variety of mangrove and brackish water communities are the main types present, growing under different conditions of tidal influence and water and soil salinity. Inland the vegetation ranges from floating and rooting aquatic herbs and herbaceous and grass swamp to swamp woodland, swamp forest, and tall luxuriant dryland forest.

In the hills and mountains changes in the structure and floristics of the natural forest vegetation are largely related to climatic differences. At low altitudes regions that experience seasonal droughts have forests and woodlands with a high proportion of deciduous trees. In general the forest becomes poorer in species and less complex in structure with increasing altitude, owing mainly to lower temperatures and increasing low cloud cover.

The boundaries of the last three main environments, which are largely forested, are not precise, as the forest types grade into one another, and the transitions themselves vary in altitude from place to place because of local effects of climate and topography. The 1000 m contour is fairly generally accepted in Malaya and Indonesia as a rough average for the boundary between the lowland and lower montane zones. The 3000 m boundary between the lower and upper montane zones is more arbitrary, but is used because above 3000 m, in the region of the highest ridges, peaks, and plateaus, the forest tends to be less variable than below this altitude, owing to the uniformly harsh climatic conditions and generally steep slopes.

The vegetation is divided into the following structural forms: forest, woodland, scrub, herbaceous vegetation, grassland, savanna. As these terms are subject to misinterpretation they are briefly defined here. *Forest* is a woody vegetation formed by trees that have touching, interlocking, or overlapping crowns. *Woodland* is an open stand of trees with a lower layer mainly of shrubs. *Scrub* consists of dense shrubs or low trees, some of which may emerge above the general canopy. *Herbaceous vegetation* is dominated by herbs, sedges, and ferns; grasses are usually present but are not conspicuous. *Grassland* is dominated by grasses and grasslike plants, although widely scattered shrubs and trees are usually present. *Savanna* consists of grassland with an open shrub or tree layer. The structural forms grade into one another and hence are not sharply distinguished either on the ground or by these definitions. The terms are used simply as a convenient communication tool.

Method of Description

The vegetation types are described for each major environment (Table 2.1). Those present in more than one major environment are discussed where they are most common. For instance, eucalypt savanna occurs from sea level to over 1500 m on both flat and hilly terrain, but is described in the section on foothills and mountains below 1000 m, as this is the zone in which it is most common.

The vegetation types of the beach, saline swamp, and fresh water swamp environments are to a large extent seral, and are in most cases described in order of increasing structural complexity. In the other major environments successional development is less important, and forest, the generally stable and dominant vegetation, is described first. Ecological features are stressed throughout. The various environmental factors interact and counteract, and the relationships between vegetation and habitat are usually complex.

Table 2.1 Major environments and vegetation types

Major environment	Vegetation type
Beach ridges and flats	Herbaceous beach vegetation Beach scrub Beach woodland *Casuarina* forest Mixed littoral forest
Saline and brackish swamps	Mangrove scrub Low mangrove forest Mature mangrove forest *Avicennia* scrub and woodland *Excoecaria* scrub and woodland *Sporobolus* grassland Nipa palm woodland
Lowland fresh water swamps	Aquatic vegetation Herbaceous swamp vegetation *Leersia* grass swamp *Saccharum-Phragmites* grass swamp *Pseudoraphis* grass swamp Mixed swamp savanna *Melaleuca* swamp savanna Mixed swamp woodland Sago swamp woodland Pandan swamp woodland Mixed swamp forest *Campnosperma* swamp forest *Terminalia* swamp forest *Melaleuca* swamp forest
Lowland alluvial plains and fans	Mixed alluvium forest Dry evergreen forest Mixed savanna *Melaleuca* savanna *Sinoga* scrub Herbaceous fern vegetation *Saccharum-Imperata* grassland *Ischaemum-Themeda* grassland *Schoenus-Eriachne* sedge-grassland
Hills and low mountains	Mixed hill forest Dipterocarp forest *Casuarina* forest *Araucaria* forest *Themeda* grassland *Imperata* grassland *Heteropogon* grassland Eucalypt savanna

Lower montane zone	Mixed lower montane forest
	Castanopsis forest
	Nothofagus forest
	Coniferous forest
	Mid-height lower montane grassland above 2500 m
	Miscanthus grassland
	Tree fern savanna
	Sedge-grass swamp
	Phragmites grass swamp
	Swamp forest
	Palm forest
Upper montane zone	Forest
	Grassland
	Herbaceous swamp vegetation

Nevertheless in some cases the effects of environmental factors are considered separately, to avoid undue complexity.

The relationships between vegetation and habitat are not always obvious. One reason is that many plant communities are in a stage of succession and, as the full range of such stages is rarely complete at any one locality, successional sequences in time have to be pieced together from several localities. Another is that too little is known about plant species, both botanically and ecologically; many of the common and well known species have a very wide range of habitat, so that their mere presence or absence provides insufficient environmental information. However, the relative abundance of such species is usually a fairly reliable habitat indicator and can be used instead of presence or absence. Sago palm, for example, grows in fresh water and brackish swamps, hence the presence or absence of sago in itself does not indicate the nature of the environment, though sites where sago grows in dense stands are undoubtedly fresh.

Man-made vegetation comprises gardens, plantations, garden regrowth, and various stages of regenerating forest, but not grasslands and savannas, which, though also largely man-induced, are well established and relatively stable vegetation types and are described separately.

The terms 'tall', 'mid-height' and 'low' refer respectively to heights of over 30 m, 20-30 m, and less than 20 m for forest, woodland and savanna canopies, and for individual trees. For grasses and herbaceous vegetation the same terms refer to heights of over 1.5 m, 0.5-1.5 m, and less than 0.5 m.

The Beach Ridges and Flats

Vegetation and habitat

Where the coast is being built up outwards by sand, a succession of ridges develops, aligned parallel to the beach. The ridges are up to 2 m high close to the coast but inland commonly merge into gently undulating flats. In places linear swampy depressions termed swales occur between adjacent ridges. The sandy soils are well drained but, owing to lack of surface run-off, low-lying parts become inundated during the wet season. The vegetation of the beach ridges and flats ranges from pioneering herbaceous communities nearest the present beach to tall mixed forest inland. Various types of swamp vegetation are found in the swales.

Vegetation types

Herbaceous beach vegetation. The first ridge behind the beach is colonised by a herbaceous community from just above high-water mark (Plate 1). This is the so-called 'pes-caprae formation', and consists mainly of low sand-binding herbs, grasses, and sedges,

Plate 1
Ipomoea pes-caprae and other creeping herbs, sedges and grasses, backed by *Casuarina equisetifolia*, colonise the first beach ridge behind the sea shore from just above high-water mark. *Avicennia marina*, in middle ground, pioneers on an area within tidal reach.

most of which root from the nodes of long runners. The creepers *Ipomoea pes-caprae* and *Canavalia maritima* usually dominate on the seaward slope, whereas grasses and sedges such as *Ischaemum muticum*, *Fimbristylis* spp., and *Cyperus pedunculatus* (*Remirea maritima*) are more prominent on the crest. Commonly the lauraceous parasite *Cassytha filiformis* overgrows a variety of host plants. A similar vegetation is found above high-water mark on the edges of coral islands.

The littoral pioneers are deep-rooting and tolerant of salt, wind, and high soil temperatures. Most have floating seeds (van Steenis 1965), a quality that probably accounts for their wide range of distribution in the Malesian region and along the north coast of Australia (Rand and Brass 1940).

Many plants that are always present in the littoral fringe also occur outside the beach environment. For instance, *Ipomoea pes-caprae* is reported from inland localities in Java (van Steenis 1965) and fresh water beaches in Nicaragua (Taylor 1963). In New Guinea *Ischaemum muticum* has been recorded at up to 300 m altitude in grassland on islands, and *Cassytha filiformis* is a common herb in the ground layer of inland eucalypt savanna.

Beach scrub. Shrubs and low trees are commonly scattered in the inner part of the herbaceous beach vegetation, but further inland often form dense scrub. This usually comprises *Hibiscus tiliaceus* and *Desmodium umbellatum* and in places is densely tangled by climbers such as *Flagellaria indica*. Some coral islands are completely covered with a low scrub, growing on almost bare coral. This scrub commonly includes *Pemphis acidula, Allophylus cobbe, Messerschmidia (Tournefortia),* and *Scaevola,* and various herbaceous and woody creepers and climbers.

Beach woodland. Elsewhere the pioneering herbaceous vegetation grades into woodland of the 'Barringtonia formation', named after one of its characteristic members, *Barringtonia asiatica*. On coasts that are being actively eroded, where the pes-caprae formation is lacking, the Barringtonia formation may directly border the beach. Continuous salt spray is probably more important than type of soil in determining the presence or absence of the Barringtonia formation, which occurs on sandy as well as on rocky coasts and also on pumiceous lava flows where these reach the sea.

The Barringtonia formation is made up of wide-crowned, low trees of *B. asiatica* and deciduous *Terminalia catappa*, gnarled, low-branched *Calophyllum inophyllum* and the screw palms *Pandanus tectorius* and, especially on coral, *P. dubius*. The ground layer is sparse where the canopy is fairly dense, but in open stands commonly consists of a closed cover of ferns, grasses, gingers, and herbs including the liliaceous *Crinum asiaticum*. On low-lying ground bordering mangrove vegetation the fern *Acrostichum aureum* and the shrub-like *Acanthus ilicifolius* feature in the undergrowth; both species, though characteristic of brackish coastal environments, have also been recorded from inland localities.

Casuarina forest. Another common pioneer in beach ridges and flats environments is *Casuarina equisetifolia* (Plate 1), which forms dense stands above high-water mark on sandy coasts, river mouths, and off-shore bars. As such stands become less dense they are undergrown and gradually replaced by various broad-leaved trees which form a mixed forest or woodland. Some very large, emergent casuarinas may persist as remnants of the former forest.

Mixed littoral forest. Mixed forest covers much of the inland beach ridges and flats. The terrain here has a very subdued relief often hardly apparent on the ground, but the striped pattern of ridges and flats shows clearly from the air because of differences in height, composition and colour of the forest.

Littoral forest is characterised by abundant palms in the shrub and lower tree layers, and by many trees having a poor stem form. Climbers, except rattan, are usually plentiful and in many places tree trunks are covered by the climbing fern *Stenochlaena*. The forest is very mixed and tall under favourable conditions — never-flooded terrain and adequate rainfall — but is poorer and lower on sites that are seasonally inundated. Common canopy trees throughout are *Syzygium* spp., and deciduous *Pterocarpus indicus* and *Terminalia* spp. Prominent on seasonally wet sites are *Planchonia papuana*, the foliage of which turns reddish immediately before a brief deciduous period, and *Melaleuca cajuputi*, which forms a smooth, small-crowned, greyish canopy. In areas that have a low and markedly seasonal rainfall the forest has a very low canopy above which trees such as *Melaleuca, Acacia*, and *Erythrina* emerge locally.

The terrain covered by littoral forest is usually traversed by more or less permanently swampy depressions aligned parallel to the ridges and termed swales. These swales have floating and submerged water plants in their deepest parts, and swamp grasses, reeds, sago palm, pandans, and swamp woodland in progressively shallower water. Mangroves are present in brackish swales within tidal reach.

A low mixed littoral forest containing trees such as *Manilkara, Diospyros, Myristica* and *Rhus* may cover the central parts of coral islands. On some large atolls a fringe of littoral forest surrounds an inner lower-lying swampy area containing pandans and pitcher plants.

Man-made vegetation

Owing to a relatively dense coastal population much of the vegetation of beach ridges and flats and coral islands is either strongly disturbed, or secondary after gardening, or replaced by coconut plantations (Plate 2). The abundance locally of *Imperata cylindrica*

Plate 2
Planted coconut (top left and right) commonly replaces the natural vegetation on beach ridges. Nipa palm (left of centre) and reed (*Phragmites karka*) border brackish tidal creek. Small crab mound in foreground.

among the grasses of the first beach ridge is also attributed to human interference.

The Saline and Brackish Swamps

Mangroves and their habitat

Mangroves are the characteristic vegetation of the tidal coastal zone. They range from pioneering low scrub on the seaward side to tall forest inland, each type being closely correlated with conditions of soil, topography, climate, tidal regime, salinity of the water, and soil drainage and aeration. Hence mangroves are commonly zoned into different communities, each occupying its own ecological niche and dominated by one or only a few species. Strips of dead and dying trees, which are a fairly common occurrence within mangrove vegetation, are another indication of the sensitivity of mangroves to environmental factors. They are attributed to sudden changes in tidal regime such as may be caused by the disappearance of a protecting off-shore bar. Some of the effects of these factors are discussed below.

Soil and topography. Mangrove vegetation is best developed in estuaries of rivers carrying large quantities of fine sediment, and along gradually shelving coast lines, protected from strong currents and wave action. Hence mangroves are more common along the south than along the north coast (Fig. 2.1). Mangroves also grow on firm clay, peat, sand, and even coral detritus, but are then not as well developed.

Mangroves follow sedimentation rather than cause it (Watson 1928; van Steenis 1965) but once established they promote the accumulation and consolidation of silt and clay and help to stabilise river banks, checking erosion by slowing down tidal and stream currents (Jones 1971; Küchler 1972). Generally they are pioneering communities preparing the habitat for the immigration of other plants (Kunkel 1966).

Climate. Mangroves are favoured by a high and evenly spread rainfall as this prevents the accumulation of salt in the soil. In seasonally dry areas mangroves tend to be less well developed, particularly where a low annual rainfall in the hinterland results in an inadequate supply of river silt. Very unfavourable for plant growth are inner tidal flats at about high water mark, where evaporation in the dry months concentrates the salt in the upper soil levels, and rainfall in the wet season is insufficient to wash the salt away (Macnae 1966). In extreme cases the central parts of such flats are bare salt deserts, as in many places along the central south coast.

Tidal regime, water salinity, and soil drainage. Mangroves are tallest along the borders of tidal creeks where the ground-water level rises and falls daily with the tides and the soil is aerated regularly. Away from creeks their height drops rapidly, and in permanently waterlogged centres of mud flats, where conditions are practically anaerobic (Macnae and Kalk 1962), they are dwarfed to a low scrub.

Mangroves have a number of adaptations that enable them to cope with the peculiar environment they live in, such as breathing roots, stilt roots, vivipary, etc. These adaptations and their functions are dealt with in several articles and text books (see for instance Richards 1964), and will not be further discussed here.

Opinions differ as to whether or not mangroves require sea water for satisfactory growth. Most mangrove species are probably salt-tolerant to varying degrees rather than salt-demanding (Richards 1964), although according to Barbour (1970) few studies have included the important criterion of ability to reproduce in fresh water.

Types of mangrove vegetation

Mangrove scrub. A dense scrubby vegetation of pioneering mangroves is found on the seaward side on muddy shores. Farthest seawards reach *Avicennia marina, Sonneratia caseolaris,* and, less commonly, *Ceriops tagal,* all of which have seedlings able to withstand prolonged submersion, the outermost ones emerging only around low tide. Scrub consisting of another at present unnamed species of *Sonneratia* colonises low banks and islets in the mouths of rivers and upstream to the limit of tidal influence (Plate 3). This

Plate 3
Cut-off river section being colonised by *Sonneratia* sp. Swampy alluvium forest with an open canopy in foreground.

species of *Sonneratia* often grows together, but rarely mixes, with pandans and *Phragmites*.

Low mangrove forest. Dense, even-aged, one-layered forest of *Rhizophora* pioneers in sheltered positions or develops after *Rhizophora* has invaded colonising stands of *Avicennia* and *Sonneratia*. In the latter case scattered wide-crowned relic trees of the early pioneers commonly emerge above the canopy of *Rhizophora*. The following stage is a mixed forest of *Rhizophora, Bruguiera,* and other mangrove species.

As pioneering mangrove forest usually has a dense canopy, its undergrowth is very sparse, and climbers and epiphytes are absent (Plate 4). As stands grow older, a ground layer of mangrove seedlings and tufts of the fern *Acrostichum* develop (Plate 5), and the characteristic climber *Derris trifoliata* begins to appear.

Circular openings of about 0.5 ha are a common occurrrence in young and middle-aged *Rhizophora* stands. From the air they look like bomb craters; on the ground they appear to be formed by dead and dying trees and to have pyramidal groups of dense saplings and seedlings mainly of *Rhizophora* in their centres (Paijmans 1966). Damage by lightning is a likely explanation for the presence of the holes. According to Anderson (1964) lightning damage to forest in general could be more frequent than is commonly thought, particularly in uniform stands with an even canopy.

Plate 4
Mangrove forest, mainly of *Rhizophora*, lining saline tidal creek.

Plate 5
Mangrove forest, mainly of *Bruguiera*, with a ground layer of *Acrostichum aureum*.

Mature mangrove forest. Mangrove forest inland from young stands has a more open canopy which admits sufficient light for an undergrowth to develop, consisting of mangrove seedlings, saplings, *Acrostichum, Acanthus,* and an occasional nipa palm. Thin woody climbers and epiphytes are common in places and include *Hoya, Dischidia,* orchids, ferns, and the ant house plant *Myrmecodia.* The *Bruguiera* trees have buttresses, and *Rhizophora* has high and wide, arched and multi-branched stilt roots as well as adventitious roots that hang down from the crown. In exceptional cases the mangrove forest is over 30 m tall with trees over 1.5 m in girth.

As the habitat becomes less saline inland the forest becomes more complex and has a more uneven and often more open canopy in which *Rhizophora* and *Bruguiera* are joined by trees such as *Camptostemon schultzii, Heritiera littoralis, Xylocarpus moluccanus, Intsia bijuga* with aerial roots and pneumatophores, and locally gregarious *Sapium indicum* with numerous low, knobbly knee roots. Shrubs and low trees of *Aegiceras corniculatum, Brownlowia argentata, Dolichandrone spathacea,* stilt-rooted *Myristica hollrungii,* palms, and pandans form a multi-levelled undergrowth. Open stands may have a dense layer of shrub pandans and tall sedges. As the proportion of fresh water species increases, the mangroves no longer regenerate and eventually disappear.

On broad, almost level margins of estuaries the brackish transition zone between saline and fresh water environments may be up to half a kilometre wide, and the mangroves grade inland almost imperceptibly into fresh water swamp forest or swamp woodland. Where the gradient is steeper, only a narrow transition zone of swamp grasses and open swamp woodland may be present between mangrove forest and dryland alluvium forest.

Avicennia scrub and woodland. In areas experiencing a low and markedly seasonal rainfall *Avicennia marina* is the most common mangrove species. It forms scrub and woodland on sandy as well as muddy soils on both the seaward and landward side of mangrove vegetation (Plate 6). On the seaward side it is the pioneering forefront of the mangroves. On the landward side low scrubby *A. marina* is in many places the last woody mangrove to persist on increasingly saline sites, separated from the bare centres of salt flats by a sparse herbaceous fringe of the succulent and edible though salty *Sesuvium portulacastrum* and some *Tecticornia cinerea.*

A. marina is widespread along the coasts of Australia, Asia, and East Africa. In west Australia it has been found some 40 km inland from the coast, lining a salt creek where conditions were probably estuarine thousands of years ago (Beard 1967). The species is easily recognised by its pale green leaves and disc-shaped groups of peg-like pneumatophores which in older trees are very numerous and extend to well beyond the spread of the crown. In experiments with mangroves Gessner (1967) found, rather surprisingly, that removal of all pneumatophores from *Avicennia* trees caused no visible harm, even after many months.

Mangrove species associated with *A. marina* are *Lumnitzera racemosa, Ceriops tagal,* and, in south-west Papua New Guinea, *Batis argillicola.* They form either a pure scrub, commonly fringing salt flats, or a low layer under emerging *A. marina.* Where tidal

Plate 6
Woodland of *Avicennia* is typical of upper tidal flats in areas of low and seasonal rainfall. Numerous pencil-like pneumatophores form most of the ground cover.

flooding is relatively frequent there is no ground layer except for the innumerable pneumatophores of *Avicennia*. Where tidal flooding is infrequent *Sesuvium portulacastrum* and some sedges form a sparse ground cover.

Excoecaria scrub and woodland. *Excoecaria agallocha* (milkwood) is characteristic of brackish fluctuating swamps on the inner side of mangrove vegetation in low rainfall areas, and forms dense scrub, or open woodland, or occurs as scattered trees emerging above a lower layer of *Hibiscus tiliaceus* and *Pluchea indica*. Its density probably depends on the duration and degree of swampiness and salinity. As fresh water influence increases, *E. agallocha* becomes mixed with trees such as *Acacia* and *Melaleuca*, and in the ground cover *Acrostichum* is replaced by grasses and sedges. Bunches of adventitious roots sprouting from the bases of *Excoecaria* and *Melaleuca* trees are probably a response to intermittent flooding.

Sporobolus grassland. In low rainfall areas a sward of *Sporobolus virginicus* (sand couch, salt couch grass) often occupies the transition zone between *Avicennia* scrub and dryland grassland or eucalypt savanna inland. The sward is dense and lush on the lowest part of the zone where it borders mangrove, but is open and lower on the landward side. The habitat is probably seasonally flooded by fresh water but is only rarely reached by the tidal salt water.

Nipa palm woodland. Woodland consisting of nipa palm (*Nypa fruticans*) covers extensive low-lying areas in estuaries subject to daily brackish flooding (Plate 7), and also lines tidal creeks where fresh and salt water meet and mix. On the lowest-lying sites nipa palm forms pure stands which have a closed canopy and no

Plate 7
Dense nipa palm vegetation in estuary of Purari River. Scattered mangrove trees emerge above the canopy.

undergrowth except for a few palm seedlings and tufts of *Acrostichum* and *Crinum*. On slightly higher ground mangroves and trees of the brackish transition zone emerge above the palms. Although stools of nipa are often seen drifting seaward with the current of a river in flood, the palm probably does much to stabilise muddy banks of estuaries through its extensive and deep root system.

Man-made vegetation

In places the mangrove vegetation has been cleared for coconut plantations and gardens. Coconuts are planted just above high-water mark wherever underlying sandy beach ridges surface. Gardens are made on 'crab islands' after the surrounding nipa has been cleared and the mangrove trees killed by ringbarking to admit more light. Crabs favouring a brackish environment build their mounds to well above high-water mark, and where crabs are plentiful some of the mounds become interconnected to form small islands. Many kinds of food crops including banana, sweet potato, pineapple, sugarcane, cassava, coconut, and breadfruit are planted and flourish on the islands, while nipa and various mangrove species persist in the tidal channels separating them.

Resources

The mangroves of Papua New Guinea are little used. A number of mangrove species have been found to be quite suitable for kraft pulp (von Koeppen and Cohen 1955; Phillips and Watson 1959), but to date none has been commercially exploited. Large areas of nipa palm along the western south coast are a potential, relatively cheap source of sugar and alcohol, both of which can be processed from the sap of the immature inflorescence. Nipa palm leaves are used by the indigenes for thatching.

Mangroves play an important role in shore protection and natural land reclamation and harbour many kinds of fish, birds and other wildlife dependent on them for survival. Although the mangroves of Papua New Guinea are in no immediate danger of excessive damage there is a need for conservation, because once destroyed, mangroves are not readily re-established (Anon. 1972).

The Lowland Fresh Water Swamps

Swamp vegetation and its habitat

The vegetation of the lowland fresh water swamps forms a continuous sequence from open water to tall mixed swamp forest, depending on the depth and quality of the water, and drainage and flooding conditions. In relatively deep water in lakes, lagoons, oxbows, blocked channels, and sluggish rivers plant growth begins with communities of free-floating aquatics. As the water becomes less deep, rooting water plants are able to establish themselves. Next in the sequence are herbaceous communities of mainly sedges, herbs, and ferns in stagnant water, while grasses predominate in swamps with moving water. Swamp grasses densely cover vast stretches of alluvial plains subject either to frequent severe and deep flooding, as along the middle courses of the Fly and Sepik Rivers, or to prolonged shallow flooding, as on some alluvial plains in southwest Papua New Guinea, and are also prominent in lakes and lagoons with slowly moving water.

In shallower swamp shrubs and trees appear and form various savanna and woodland communities widely differing in height and density. Swamp forest is the last, and in its mixed form the most complex, member of the sequence. It is the dominant vegetation type in the mainly wooded back swamps of the upper and lower courses of the larger rivers, while scattered patches of it occur within grass swamps on submerged hill spurs, levees, and meander scrolls of old river courses (Plate 25). In transitions between swamp forest and swamp woodland, the forest tends to occupy the better aerated and better drained sites. It is tallest in strongly fluctuating but permanent swamps just behind river levees, while in quiet swamps further away from the river it is lower, thinner stemmed, and usually more open, though it is occasionally denser. On the back slopes and tops of levees swamp forest grades into a swampy type of alluvium forest as flooding becomes less prolonged.

Types of swamp vegetation

Aquatic vegetation. This type consists of small free-floating aquatics such as *Lemna, Azolla imbricata, Pistia stratiotes, Utricularia,* and minute sedges. These either form a mixture, or grow in a mosaic in which some members of the community form single-species colonies.

From the air these colonies can be readily identified by their different colours; patches of *Azolla*, for example, show a reddish brown colour, while those of *Pistia* are yellowish green.

From depths of less than about 3 m rooting water plants are able to establish themselves. They include submerged aquatics such as *Ceratophyllum*, semi-submerged, white and blue-flowering water-lilies of the genera *Nymphaea* and *Nymphoides*, and the large pink-flowering lotus *Nelumbo nucifera*. They occupy the shallowing margin between open water and grass swamp, and in places cover entire lakes and lagoons that have a uniform depth.

Herbaceous swamp vegetation. Herbaceous communities consisting of sedges, herbs, and ferns are characteristic of stagnant, permanent, relatively deep swamps (Plates 8 and 9). They root in a partly floating mat of waterlogged peat and organic debris, and grow up to 2.5 m above the level of the often dark-coloured water. Common members are the coarse, robust sedges *Thoracostachyum sumatranum* and *Scleria* sp., the tall, fleshy, broad-leaved herb *Hanguana malayana,* and the fern *Cyclosorus.* Local facies are often present and in them herbs, sedges, or ferns predominate. In monsoonal areas a mosaic pattern of local dominance is often present; this may be a result of irregular burning. Elsewhere the mosaic pattern is probably related to the depth and nature of the water, and to chance distribution.

The centre of a herbaceous swamp usually has a lower and poorer vegetation, often consisting of only a few genera such as the fern *Gleichenia,* the club moss *Lycopodium*, and low sedges. Towards land, grasses become more prominent, and scattered shrubs, pandans, low sago palms, and climber-covered small trees grow on

Plate 8
Herbaceous swamp vegetation of mainly ferns and sedges grading into swamp savanna of low climber-covered trees on slightly higher ground.

Plate 9
Sepik River area. Tall *Scleria* and clumps of stunted sago fill a swampy depression within *Themeda-Ischaemum* grassland.

low hummocky rises. *Phragmites karka* often dominates along gently sloping swamp margins, whereas *Pseudoraphis spinescens* and *Ischaemum polystachyum* commonly form narrow bands along more steeply sloping, wet-dry margins. In north-eastern Papua, Taylor (1959) noted that herbaceous swamp vegetation occurred on peat deeper than 2 m and grew taller in areas without peat or with shallower peat.

Leersia grass swamp. Mid-height swamp grasses such as *Leersia hexandra* (rice grass), *Echinochloa stagnina, Oryza* spp. (wild rice), *Panicum* sp., and *Hymenachne amplexicaulis* occupy permanently swampy parts of river plains that are under up to 3 m of water in the flood season (Plate 10). During this time 'islands' of swamp grass come adrift and float downstream. They often become anchored again on the way, but occasionally reach the sea.

In lakes and lagoons the same grasses fill shallow embayments, surround islets, and mark submerged levees while *Ischaemum polystachyum* often occupies wet-dry fringes. The grass stems trail along the top of the water but their culms rise to over 1 m above it. Herbs such as *Polygonum, Ludwigia,* and *Ipomoea aquatica* are anchored in the grass mat and reach out over open water.

Saccharum-Phragmites grass swamp. Tall swamp grasses, mainly *Saccharum robustum* and *Phragmites karka,* grow in swamps that are shallower than those occupied by mid-height swamp grasses, and may be intermittently dry. *S. robustum* commonly forms pure stands on relatively high river levees subject to frequent but only brief flooding (Plate 11). *Phragmites karka, Coix lachryma-jobi* and, particularly in the Sepik flood plain, *C. gigantea* are prominent on low levees and levee back slopes, with *P. karka* increasing

Plate 10
Islands of swamp grass lining flooded Aramia River. Planted bamboo grows on low hills above flood level.

Plate 11
Tall dense *Saccharum robustum* on levee of Strickland River.

in abundance down slope. Shrubs and low trees of *Glochidion, Nauclea coadunata, Mitragyna speciosa,* the palm *Livistona* and, in the Fly River area, *Barringtonia tetraptera* are widely scattered within this community. The habitat is marginal for most woody vegetation; many trees are dead or dying, and most are overgrown with climbers. In the dry season, after the flood waters have fallen, a thick mat of coarse, dead and live grass stems is revealed, covering underlying peaty muck, and adventitious roots can be seen to sprout from the nodes of the ascending stems. They consist of thin, thread-like rootlets, and thicker roots which themselves also develop adventitious rootlets.

The habitat of *P. karka* overlaps that of *S. robustum* but *P. karka* grows over a wider range of conditions. *S. robustum* is a fresh water swamp grass favouring moving water. It is not likely to survive long dry periods, and does not tolerate much shade. Planted *S. robustum* has been recorded at 2300 m a.s.l. *P. karka* grows in permanent swamps and on sites that are seasonally dry for several months, in both stagnant and moving water, and in fresh water and brackish environments; it also tolerates growing under a fairly dense canopy of trees or pandans. *P. karka* occurs naturally to over 2500 m a.s.l.

Pseudoraphis grass swamp. *Pseudoraphis spinescens* (water couch grass) is a low creeping or trailing swamp grass that is most extensive in south-west Papua New Guinea. Here it forms dense, almost pure, matted swards on flood plains that are seasonally dry but are shallowly inundated probably for most of the year (Plate 12). During the short period of emergence, the grass recuperates, but it is heavily grazed by wallabies and herds of deer. Elsewhere the grass forms

Plate 12
Seasonally dry *Pseudoraphis* grassland bordering Bensbach River in south-west Papua New Guinea. Low and deep-crowned *Barringtonia* trees give the landscape a parkland aspect.

narrow belts along wet-dry swamp margins, and is also occasionally found in swales between beach ridges. It is used as a lawn grass in villages along the Sepik and Ramu Rivers (Henty 1969).

Mixed swamp savanna. Mixed swamp savanna is transitional between purely herbaceous swamp and swamp woodland and occurs in permanent more or less stagnant swamps. It consists of a dense ground layer very similar to herbaceous swamp communities and an open emergent layer of trees belonging to the limited number of genera such as *Nauclea, Campnosperma, Syzygium,* and *Melaleuca,* that are able to put up with the adverse edaphic conditions (Plate 8).

Melaleuca swamp savanna. Swamp savanna of ti-tree or paper-bark (*Melaleuca* spp.) is characteristic of fluctuating back swamps of the middle Fly and Strickland Rivers, and also occurs sporadically on low-lying beach flats along the monsoonal south and south-west coasts. *Melaleuca* trees form an even, almost pure, one-layered canopy. The stands are mostly rather open and consist of low and crooked trees, but they vary widely in density, height, and stem form. Some stands are of tall and straight trees, and resemble *Melaleuca* swamp forest (see later) but for the characteristic ground layer present — usually the grass *Phragmites* either pure or mixed with lower swamp grasses and coarse sedges.

Several species of *Melaleuca* are present in this vegetation type, but individual stands consist predominantly of one species, the most common being *M. cajuputi* and *M. leucadendron*. Other trees, which are very sparse, include *Nauclea coadunata,* pandans, the palm *Livistona,* and, in the understorey, sago.

In the wet season *Melaleuca* swamp savanna is often deeply inundated, and floating water plants from nearby permanent open water may cover the surface. At the end of the dry season many stands have become dry and easy to penetrate. The trees are then seen to grow on hummocks, and to have fire-scorched trunks, and locally sago palms have been killed by fire. A crown fire may completely defoliate the trees without killing them. Most *Melaleuca* swamp savanna probably results from repeated fire damage to a mixed woody vegetation, though some may be either a true edaphic climax or a fire-arrested seral stage in the development towards forest.

Mixed swamp woodland. In permanent swamps the tree storey of mixed swamp woodland is generally open and ranges from low to tall. Common trees are *Campnosperma, Nauclea coadunata, Mitragyna ciliata,* and *Timonius*. Palms and pandans fill in much of the space below the trees, and *Hanguana,* tall sedges, and the fern *Cyclosorus* form a dense ground layer. Fleshy climbers and climbing ferns usually abound and include *Flagellaria, Nepenthes,* and *Stenochlaena*.

On sites that are seasonally dry swamp woodland is often low, thin-stemmed, and as dense as forest. Adventitious roots are a characteristic feature, probably developing in response to the fluctuating water table, and the ground layer is usually low and patchy. Common trees throughout are *Carallia brachiata* and species of *Syzygium, Mangifera, Garcinia,* and *Acacia*. In monsoonal south-west Papua New Guinea *Barringtonia tetraptera* is prominent in swamps that are deeply inundated in the wet season, whereas *Tristania* and *Xanthostemon* show a preference for slightly higher lying ground. In seasonally dry swamp woodland the proportion of trees with a cork-like layered bark is higher than average, probably as a result of natural selection through repeated fire damage.

Sago swamp woodland. Sago palm, *Metroxylon sagu*, is a widespread tall shrub that grows in more or less permanently swampy woodland. All gradations occur from stands of pure sago palm virtually without trees (Plate 13) to woodland with a rather dense layer of trees and an open lower tier of sago. The palm grows best in shallow swamps where there is a regular influx of fresh water. Here its fronds are up to 14 m high, and the flowering, starch-producing tree-like stems reach 20 m. The palm multiplies by forming suckers around the base of the old stems, commonly after flowering. Both smooth and spiny varieties of sago palm are present.

The ground layer varies with the density of the sago palms and the degree of swampiness. Under dense sago there is no undergrowth, the peaty soil is layered with fallen dead fronds, and the palm's numerous tiny pneumatophores form the only live 'cover'. Open stands have an undergrowth of shrub pandans, *Hanguana*, sedges, or *Phragmites* where the water table is permanently near the surface, and a ground layer of grasses, gingers, and ferns where the water table is well below the surface at least for part of the year. Sago stands become stunted in transitions to permanent herbaceous swamp, in brackish environments, and on sites where the water table temporarily sinks deep enough to cause drought stress. On

Plate 13
Sago palm woodland, flowering stems in background.

such sites the palms do not flower, but sucker strongly. On sites transitional to deep herbaceous swamp sago grows in scattered circular groves.

Pandan swamp woodland. Swamp pandans occupy a habitat similar to that of sago palm, but have a wider range. They form rather open to quite dense, pure stands up to about 8 m high in shallow, fresh to brackish, stagnant to frequently flooded swamps (Plate 14). An

Plate 14
Pandanus swamp woodland in old river bed.

upper layer of scattered trees may be present. Air photos and reports indicate that pandans cover extensive areas in the flood plains of the middle Sepik River. The species of *Pandanus* present in Papua New Guinea and in particular their ecology are incompletely known.

Mixed swamp forest. This is the most common type of swamp forest. It has a generally rather open, but occasionally dense canopy which under favourable conditions reaches a height of 30 m. The canopy has a relatively even height which gives it an appearance of uniformity from the air. Some of the trees often found in the canopy are *Campnosperma, Terminalia canaliculata, Nauclea coadunata, Syzygium, Alstonia scholaris, Bischofia javanica,* and *Palaquium,* but many more are also frequent.

The lower tree strata are open, because most trees have their crowns in the canopy. Characteristic trees below the canopy are *Alstonia spatulata, Barringtonia, Diospyros, Garcinia,* and *Gynotroches axillaris.* Sago palm and pandans commonly form a substratum which is, in places sufficiently dense to exclude all undergrowth. The density of the shrub and herb layers appears to vary with flooding and light conditions. Where flooding is prolonged and deep, there are virtually no shrubs and herbs, and the forest is easy to penetrate once the floor has become dry (Plate 15). Open stands may have a dense ground layer of *Hanguana* and tall sedges. Thin lianes, fleshy climbers, and climbing ferns often thickly cover

Plate 15
Swamp forest in fluctuating back swamp of Fly River. Large buttressed tree in centre is a *Terminalia*.

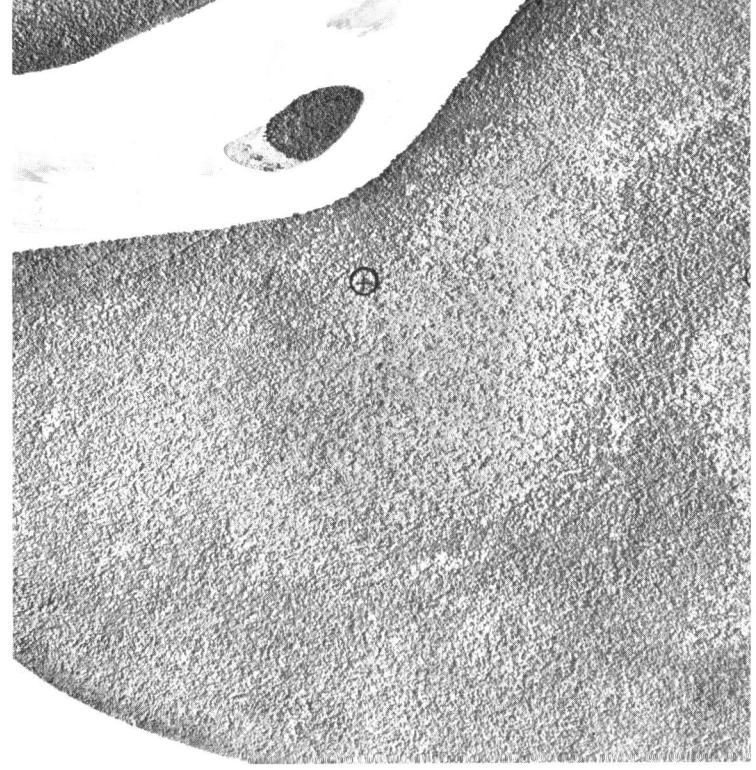

Plate 16
Campnosperma swamp forest, identifiable from the air by its grey-toned smooth canopy of flattish crowns. The canopy of the surrounding mixed swamp forest is darker toned and more granular.

tree trunks, but rattan is usually rare. Buttresses are not conspicuous, but stilt roots, sprawling surface roots, and knobbly, peg-like, and knee-shaped breathing roots abound.

A particular type of mixed swamp forest occurs in river deltas flooded by fresh water backed up by the incoming ocean tide through a maze of creeks and channels. Flooding is daily near the coast but becomes less frequent inland. The stilt-rooted *Myristica hollrungii* is a characteristic component of swamp forest in this environment. The type has many features in common with mangrove forest, such as the coastal location, the even, dark-toned, small-crowned appearance of the canopy from the air, and the commonness of stilt-rooted trees. Because of this it is sometimes referred to as 'fresh water mangrove'.

Campnosperma swamp forest. Two species of swamp *Campnosperma*, *C. brevipetiolata* and *C. coriacea*, are present in this vegetation type. *C. brevipetiolata* is widespread throughout lowland swamp forest and commonly forms pure stands, but *C. coriacea* appears to have a more limited distribution. Some stands consist of both species in about equal proportions.

The densest stands of *Campnosperma*, though not necessarily the tallest, are found in permanently flooded back swamps (Plate 16). Sago usually forms a more or less dense substratum. Buttresses, stilt roots, and knee roots abound, a thick layer of organic debris covers the forest floor, and the underlying soil is peaty. Regeneration appears to take place through irregularly spread groups of seedlings and saplings. The swamp Campnospermas are large-leaved trees developing long thick straight boles and wide flat-topped crowns. Pure stands are readily recognisable from the air by their smooth, even, grey-toned canopy, although a swamp-dwelling species of *Calophyllum* occasionally forms dense stands resembling those of *Campnosperma*.

In Papua New Guinea *C. brevipetiolata* is rarely found on dry land. In the Solomons, however, it is the most abundant big tree species on many dry sites and it forms continuous hill slope stands on a number of Micronesian islands (Whitmore 1974).

Terminalia swamp forest. This type is found mainly on Bougainville, where *Terminalia brassii* often grows together with *Campnosperma* and locally dominates in the canopy of open swamp forest. Low-lying, frequently flooded, bouldery and sandy river tracts and peat swamps with flowing water are the habitats. *T. brassii* grows into a tall straight-boled tree with a large girth and a wide cauliflower-shaped crown which is very distinctive from the air (Plate 17). *T. brassii* swamp forest may be a seral stage in the development towards swampy alluvium forest.

Melaleuca swamp forest. Swamp forest of *Melaleuca* spp. is mainly confined to monsoonal south-west Papua New Guinea, where it occurs in narrow bands in seasonally dry swamps along rivers and creeks (Plate 18). The main species is *M. cajuputi*, but other species or varieties of *Melaleuca* are present. The trees are slender, growing to a height of 30 m or more, but rarely reaching a girth of 1.5 m. They have white stems with paper-thin flaking bark and are closely spaced but, as they have small crowns, the canopy closure is rarely

Plate 17
Terminalia brassii swamp forest, distinctive from the air by its cauliflower-shaped crowns.

Plate 18
Melaleuca swamp forest. The interior (background) consists of almost pure *Melaleuca*. Various other tree species and many climbers and epiphytes are present along the forest edge (foreground). Most trees grow on hummocks.

more than 50 per cent. Owing to a lack of trees other than *Melaleuca* and a virtual absence of a shrub and herb layer, the forest has an aspect of monotony rare in forest of the tropics. Some tree bases have masses of adventitious roots growing up to flood level. Groups of young trees are present in places and probably assure the continuation of the type. A very similar paperbark forest is described from northern Australia (Story 1970).

The kind and frequency of associated trees and ground flora are related to the drainage conditions and length of inundation. *Acacia* spp. and *Dillenia alata* are most frequent on slightly elevated terrain within and along the edges of the swamp, and on such sites *Imperata cylindrica* may form a sparse ground cover. Prolonged inundation is indicated by the presence of the grass *Pseudoraphis* and the herbs *Philydrum lanuginosum* and *Eriocaulon*. The tree *Barringtonia tetraptera*, aquatic herbs, and the rush-like sedge *Eleocharis* are found on very low-lying, more or less permanently swampy sites.

It is not clear whether *Melaleuca* swamp forest is a disclimax replacing mixed swamp forest after repeated fire damage, or a true edaphic climax, or both. The common occurrence of charred trees indicates that dry-season fires sweep through most stands.

Plant succession

Successional development in swamps results from a decrease in depth of water through either a rise in the level of the floor or a lowering of the water table. In stagnant swamps that have no input of silt, the floor of the swamp rises when accumulation of organic debris exceeds decomposition. There is no factual information available on the rate of either, but in tropical lowland swamps decomposition of organic matter is probably not much slower than accumulation, and although peat appears to be forming in most stagnant swamps, the plant communities of these swamps can be regarded as near-stable edaphic climaxes.

On the other hand, in swamps flooded by moving, silt-loaded water, sedimentation takes place and the swamp floor rises, enabling successional development of plant communities. The first plants may colonise river levees, where sedimentation is greatest, and later spread on to the flood plains. As sedimentation continues the succession at any one site advances through stages similar to those observable at any one time in different increasingly shallow swamps. Commonly the plant cover itself promotes succession by trapping silt and so speeding up sedimentation. However, dense swamp grass may prolong swampy conditions behind a levee by blocking the existing outlets to the river.

Successional development may be either accelerated or reversed by abrupt changes in a river course and the consequent changes in flooding regime. Thus, while plant succession is speeded up or is proceeding smoothly in some parts of a flood plain, it may be temporarily set back or halted in other parts. In areas that are seasonally dry, the plant succession may be retarded by fire. In areas of grass swamp and herbaceous swamp fires damage and kill pioneering shrubs and trees and, in exceptionally dry periods, may encroach upon established woody swamp vegetation. Thus, grass swamp may be kept in a permanent subclimax state, and increase in area at the expense of wooded swamp. Fires may also alter the floristic composition.

Resources

The fresh water swamp environment is relatively little used and, apart from fire damage in seasonally dry areas, is little disturbed. Sago forms the staple food for many people in the lowlands and is often planted, both within and outside its natural range of habitat. Harvesting of sago causes little damage to the environment, as only stems about to flower are removed.

As swamp forests pose serious access problems, timber exploitation is restricted to a few gregarious species, such as *Terminalia brassii,* that have market value. *Campnosperma,* though producing a good peeling log, has not been harvested on a large scale in Papua New Guinea. *Melaleuca* has a hard and durable wood, and its leaves yield a medicinal oil. However, the tree is little used in Papua New Guinea, although it has been cultivated in several places in Malaya and Indonesia.

The Lowland Alluvial Plains and Fans

Vegetation and habitat

Forest is the natural and, up to the present day, dominant vegetation of the lowland alluvial plains and fans, although the indigenous population, through gardening and burning, has converted large areas of forest into grassland, and modern large-scale agricultural development and intensive forest exploitation are also making major inroads.

The climax forest grows on well to imperfectly drained plains and gently sloping fans that have deep soils and are either not flooded or are flooded infrequently and for short periods only. The water is well below the surface for most of the year. Minor drainage deficiencies such as are indicated by the presence of pools of standing water, which form after heavy rain and may remain for several days, appear to have little or no effect on the forest. The flat to gently undulating plains of monsoonal south-west Papua New Guinea with their vegetation of dry forest, woodland and savanna resemble northern Australia and are quite unlike lowland alluvial plains elsewhere in Papua New Guinea.

The forest

Tall and floristically and structurally very rich, mixed alluvium forest is the most luxuriant of the lowland rain forests in Papua New Guinea. It is described here in some detail, as it is used as the standard to which other forest types will be related (Plate 19).

The forest structure is irregular throughout all layers. The canopy is very variable in height, closure, and crown sizes, features that are

Plate 19
Interior of tall forest on relatively well-drained alluvium, rather open spot rich in climbers. The undergrowth is well developed and includes many tree seedlings and *Licuala* palm. Young rattan palm at bottom left.

particularly striking from the air. It is rather open and has many gaps in which only lower trees are present. It is generally about 30-35 m high, but irregularly scattered trees emerging partly or fully above the canopy may reach 50 m or more. Most emergents and many canopy trees have wide crowns, tall straight boles, and high and wide buttresses, some trees attaining a girth well over 2.5 m. Tree species constantly present in the upper storeys are *Pometia pinnata, Ficus* spp. including strangling figs, *Alstonia scholaris,* and *Terminalia* spp. The lower tree strata are usually rather open, as the number of trees below the canopy is relatively low. Some trees with stilt roots are invariably present. Typical lower storey trees are *Garcinia, Diospyros, Myristica, Maniltoa* and *Microcos.*

The cover and density of the shrub and tall herb layer vary with the amount of light penetrating through the canopy, and with the types of shrubs present. A predominance of spreading, multi-branched shrubs results in a high cover, whereas a dense layer of slender saplings may have only a low cover. Palms are a feature of the shrub layer. They consist of small tree palms, true shrub palms, and young rattan palms, and include the common *Licuala* with its conspicuous fan-shaped leaves. Tall gingers and Marantaceae locally form a dense layer. Pandans are rarely common, and tree ferns and bamboo are scarce.

The low herb layer forms an irregular and patchy ground cover. It may be almost absent, as where shrub palms are abundant, or quite dense, as where it is formed by *Selaginella, Elatostema,* Marantaceae, or Commelinaceae. Otherwise the herb layer consists mainly of ferns, tree and rattan seedlings, and some forest grasses and sedges.

Thin and thick woody lianes, fleshy epiphytic climbers, and climbing ferns are usually common. Climbing rattan is invariably present, but is dense only below canopy openings. Epiphytic ferns and orchids are plentiful in the crowns of canopy trees, especially in old trees with open crowns, thick branches, and a rough bark. Unlike those in some high mountain forests, most orchids have inconspicuous flowers.

Environmental controls

Under less favourable conditions of drainage, soil, and climate alluvium forest is less luxuriant. Drainage is the main factor, and it affects both structure and floristics of the forest. Soil mainly influences the structure, and differences in amount and distribution of the rainfall are reflected most markedly in species composition. The effects of these factors on alluvium forest are discussed in more detail below.

Drainage. Where wet-season inundation is prolonged the forest canopy is lower and smaller crowned than in forest on well drained terrain and has more large gaps (Plate 3). The emergents are more widely spaced, and in the small girth classes the number of trees is lower. Climbers, particularly rattan, are more common, and small groups of sago palms are present in the understorey. The shrub layer is denser, because there are more palms and pandans. Tall sedges and low shrub pandans form part of the ground cover. More trees have stilt roots, and some trees and loops of thick lianes that descend to near ground level develop abundant adventitious roots, presumably in response to inundation.

With more prolonged inundation the forest usually becomes broken up into groups of trees and patches of shrubbery overgrown by rattan and other climbers, but may retain its tall stature and grade into swamp forest. Certain trees that occur sparsely on better drained sites increase in numbers and tend to grow gregariously. These include *Planchonia papuana, Bischofia javanica, Terminalia complanata, Cananga odorata, Teysmanniodendron bogoriense, Intsia bijuga, Nauclea coadunata, Alstonia scholaris, Vitex cofassus* and *Anthocephalus chinensis.*

In riverside forest subject to frequent brief flooding the shrub layer is commonly very open, the herb layer is almost absent, and the leaf litter is very sparse. In places, however, a convolvulaceous creeper, never collected fertile and at present unidentified, forms a dense cover on the forest floor.

Soil. On heavy-textured soils the forest tends to be lower, smaller-crowned, and denser, probably owing to poor permeability. As such soils are usually present on back plains subject to more or less prolonged inundation, the effects of soil texture and lack of surface run-off cannot be readily separated.

Several flat to gently undulating plains and fan slopes have a high proportion of either coarse river deposits, or iron and manganese concretions in their soils, often accompanied by poor drainage. The forest on such sites is typically thin-stemmed and small-crowned, with a high overall tree density, but in places the canopy is broken up by lines of larger-crowned trees and patches of swamp vegetation. Locally predominant trees are *Intsia, Casuarina, Campnosperma, Pterocarpus* and dipterocarps.

Climate. Where there is a marked dry season the forest canopy contains a much higher proportion of deciduous and semi-deciduous trees than in regions experiencing a high and more evenly spread rainfall, where the percentage of deciduous trees is very low. Such trees are *Bombax ceiba, Anisoptera polyandra, Terminalia* spp., *Ficus* sp., *Garuga floribunda*, and many others. Deciduous trees may be common even on sites that are usually wet; their presence here is probably due to periodic water stress. Structurally, alluvium forest with deciduous trees is similar to its evergreen counterpart, except for a local abundance of climbing and scrambling bamboo.

Other vegetation types

The other vegetation types that are most common on plains are dry evergreen forest, two types of savanna, *Sinoga* scrub, herbaceous fern vegetation, and three types of grassland. The vegetation on the plains of monsoonal south-west Papua New Guinea is of special interest, as five of the eight vegetation types are restricted, or almost restricted, to this region. The vegetation communities of these plains have been previously described by Brass (1938), and by van Royen (1963a), who also studied an adjacent area of Irian Jaya north of Merauke, which has a very similar vegetation.

Dry evergreen forest. Dry evergreen forest is restricted to an area of low rainfall (1800-2500 mm) in south-west Papua New Guinea where gently undulating, well drained plains form the main habitat. It is less luxuriant than forest on plains elsewhere, and has a very different floristic composition.

The canopy, between 20 and 25 m high, is rather open, but the lower tree strata are dense. Tree trunks are rather frequently low-branched and crooked, and leaves are predominantly small. Though low palms abound in places, the shrub layer is generally fairly open and the ground cover is very sparse, so that the forest is easy to walk through. Woody lianes are common, but rattan, fleshy climbers, and epiphytes are not conspicuous. The leaf litter is thin, and very dry in the dry season. Near the border with Irian Jaya a dense understorey of tall bamboo is commonly present, virtually excluding all other undergrowth, and where the forest canopy opens up the forest grades into bamboo woodland.

The commonest trees in dry evergreen forest are *Acacia* spp., the myrtle family with *Tristania, Syzygium, Rhodamnia,* and *Xanthostemon,* Proteaceae with *Oreocallis* and *Grevillea,* Rutaceae with *Halfordia* and *Flindersia,* and the genera *Maranthes* (syn. *Parinari*) and *Mangifera*. With increasing rainfall the forest loses its characteristic trees and grades into normal rain forest.

Fires from the surrounding savanna often enter the forest, killing shrubs and fire-sensitive trees. Gingers, other tall herbs, and forest seedlings spring up, and if there are no further fires the vegetation reverts to forest. If burning occurs frequently, however, grasses become established and increase the fire hazard. Only forest plants able to endure repeated fires survive and become predominant, and with the invasion of species from the surrounding savanna the erstwhile forest turns into savanna.

Mixed savanna. This type of savanna, like dry everygreen forest, is restricted to monsoonal south-west Papua New Guinea. Its structure and floristics vary with the local relief and drainage conditions, and with the frequency of burning. On well drained, gently undulating terrain, it is as tall as the dry evergreen forest it replaces after repeated fire damage, but it is much poorer in species (Plate 20). Trees are irregularly spaced but moderately dense overall. On poorly drained flat terrain mixed savanna is lower and more open, and with increasing wetness of the terrain passes through a transitional savanna with thin-stemmed crooked trees into sedge-grassland.

The commonest trees are species of *Tristania, Melaleuca, Acacia,* and *Xanthostemon*. Eucalypts are present in many places, but never in abundance. Shrubs, including *Acacia leptocarpa, Choriceras tricorne,* and *Helicteres angustifolia,* are also relatively tall and dense, and compete with the grasses *Imperata cylindrica, Ophiuros tongcalingii,* and *Ischaemum barbatum* which form most of the ground layer. *Melaleuca symphyocarpa, M. viridiflora, Banksia dentata,* and *Grevillea glauca* become increasingly common as drainage conditions deteriorate, and often are the only tree species in the transition zone to sedge-grassland.

Various stages of tall mixed savanna are present, reflecting the frequency of fires, and ranging from a 'grassland' community with relatively few trees and shrubs to a rather dense 'woodland' with

Plate 20
Tall mixed savanna on relatively well-drained, gently undulating terrain in south-west Papua New Guinea. Common trees are *Melaleuca*, with smooth light-coloured trunks, and *Tristania*, with fissured darker coloured bark. The grass layer is mainly *Imperata cylindrica*.

both shrubs and grasses important in the ground cover.

Melaleuca savanna. *Melaleuca* savanna is found mainly in monsoonal regions but also in areas with a relatively high and evenly spread rainfall. It is extensive in south-west Papua New Guinea, where it is established on low-lying seasonally inundated flats bordering the coastal plain (Plate 21). Elsewhere it occurs sporadically, up to about 500 m altitude, on seasonally waterlogged plains and fans and also on permanently dry hill slopes. It is present on some of the islands off the eastern tip of mainland New Guinea where the annual rainfall in places is over 3000 mm (Brass 1959: 64). Most if not all of the areas now under *Melaleuca* savanna were probably originally covered by mixed woody communities.

As in *Melaleuca* swamp savanna, *Melaleuca* forms pure stands, usually of thin-stemmed, crooked trees. The commonest species in south-west Papua New Guinea is *M. viridiflora,* while elsewhere *M. cajuputi* and *M. leucadendron* are predominant. The ground layer varies regionally and with local relief and soil water conditions. On the wet-dry plains in south-west Papua New Guinea the ground cover consists mainly of sedges and grasses such as *Germainia capitata* and *Ischaemum barbatum.* Tall grasses dominate on plains and fans that have adequate soil water in the dry season, and mixtures of *Themeda australis* and *Imperata cylindrica* form the ground layer on hill slopes.

Melaleuca savanna and the other monsoonal communities of south-west Papua New Guinea have strong structural and floristic affinities with those of northern Australia. For instance, *Melaleuca viridiflora* savanna (Story 1970), the *Tristania-Grevillea-Banksia*

Plate 21
Melaleuca savanna on seasonally inundated plain in south-west Papua New Guinea. Drainage lines are marked by a mixed forest or woodland of taller darker toned trees.

community described and depicted by Christian and Stewart (1953), and low heath (Pedley and Isbell 1971) of northern Australia have their almost identical counterparts in New Guinea. In both regions Myrtaceae are important savanna components, but eucalypts are much more common in northern Australia. Very similar conditions prevail in some South American savanna regions, where, as in New Guinea, the main vegetational contrast between wooded and herbaceous savanna is attributed to soil water variations (Eden 1970). However, in the savannas of Guyana there does not appear to be evidence of recent large-scale retreat of forest as a result of burning, and Eden suggests that the present extensive savannas could have become established during a period of drier climate in the Pleistocene.

Sinoga scrub. The woody *Sinoga lysicephala* forms a low scrub in south-west Papua New Guinea in mosaic with *Schoenus-Eriachne* sedge-grassland (see later). This scrub is very dense in many places but opens up where it grades into sedge-grassland. It is killed back regularly by fire and is mostly less than 1 m high, but where it has not been burnt for several years it can reach 5 m. *Sinoga* also occurs singly as a tall shrub in mixed savanna.

Herbaceous fern vegetation. On the plains along the middle reaches of the Sepik River herbaceous vegetation dominated by *Dicranopteris linearis* locally forms a dense cover within *Ischaemum-Themeda* grassland. *D. linearis* is in places over 1.5 m high, emerging above a ground cover of sedges, grasses, *Lycopodium*, and climbing *Lygodium* and other ferns. The ground layer varies in density inversely with the layer of *D. linearis*. The type forms a mosaic with *Ischaemum-Themeda* grassland in the eastern part of the Sepik plains but dominates completely over large tracts further west (Heyligers 1972a). This is probably due to the lower rainfall and hence more effective dry-season burning in the east than in the west. *Dicranopteris* appears to be replaced by grasses where fires are frequent, and to remain dominant where burning is irregular.

Herbaceous fern vegetation dominated by *Dicranopteris, Gleichenia,* or *Pteridium* is also present in the foothills and low mountains zone, both within grassland near forest borders and as isolated patches within forest.

Saccharum-Imperata grassland. This vegetation type consisting of tall grassland dominated by *Saccharum spontaneum* (pit-pit) and *Imperata cylindrica* (kunai), covers large areas on alluvial plains and fans mainly in regions which have a low and markedly seasonal rainfall (Plate 22). The habitat is imperfectly to well drained, and soil moisture conditions are relatively favourable throughout the year.

S. spontaneum grows to a height of 3.5 m and *I. cylindrica* to over 1.5 m. Common associated grasses are mid-height *Sorghum nitidum,* tall *Ophiuros tongcalingii* and *Coelorhachis rottboellioides* which have culms to over 2 m, and tall but straggling *Apluda mutica.* After burning, *Imperata* is the first to sprout afresh from its underground rhizomes and temporarily dominates the new sward. It is also tolerant of temporary waterlogging, in contrast to the other

Plate 22
Tall *Saccharum spontaneum-Imperata cylindrica* grassland on plain subject to shallow periodic flooding.

tall grasses, and forms permanent, pure and very dense stands on low-lying, seasonally inundated beach plains.

Herbs are sparse, owing to the dense shade cast by the tall grasses. However, low perennials, such as *Pygmaeopremna sessilifolia* and *Curculigo orchioides,* become conspicuous shortly after burning, when they flower and fruit during a brief period of reduced grass cover. Fire-tolerant low trees and shrubs of *Nauclea coadunata, Antidesma ghaesembilla, Albizia procera,* and pandans are widely scattered throughout and give the grassland a savanna aspect in places where they are densest, as in depressions and along seepages.

Though most extensive on plains and fans, *Saccharum-Imperata* grassland also occurs within mid-height grassland on hilly terrain wherever moisture conditions are favourable, as along drainage lines and forest borders, and on footslopes. There can be little doubt that *Saccharum-Imperata* grassland is a fire-disclimax replacing forest after repeated gardening and burning. Its ecological status is the same as that of grassland on hilly terrain, and is further discussed in the section on the foothills and low mountains.

Ischaemum-Themeda grassland. This grassland type, dominated by *Ischaemum barbatum* and *Themeda australis* (kangaroo grass) covers much of the vast plains along the middle reaches of the Sepik River between the grass swamps of the river flood-plain proper to the south, and taller grassland and forest on hilly terrain to the north (Plate 9). It is also present in the valleys of the Markham and Ramu Rivers. These inland plains lie between coastal mountain ranges to the north and the central cordillera of mainland New Guinea to the south, and as a result have a relatively low and seasonal rainfall which in the Sepik valley ranges from about 1800 mm in the east to

2500 mm in the west. The grassland is developed on strongly weathered, leached and acid heavy clays which are subject to alternate drying and waterlogging.

Over large areas the two main grasses grow in a mixture, the drought-resistant *Themeda* rooting superficially, and *Ischaemum* penetrating into the wet subsoil. Where the local relief permits surface run-off, the two species are segregated (Reiner and Robbins 1964), *Themeda* commonly being associated with *Capillipedium parviflorum* and *Alloteropsis semialata* on the drier terrain, and *Ischaemum* growing together with *Apluda mutica* on the wetter sites. Large stretches of grassland are treeless, but widely scattered shrubs are present everywhere, and woody vegetation borders the middle and lower courses of the sparse creeks crossing the grassland.

The origin of the *Ischaemum-Themeda* grasslands is uncertain. Reiner and Robbins (1964) suggest that they are an old and stable disclimax resulting from the clearing of forest and subsequent burning and gardening by people migrating northwards, and that tall mixed grassland further north is relatively dynamic and indicates a shorter period of biotic interference. Haantjens *et al.* (1965) accept the man-made origin of the two types of grassland through shifting cultivation, but stress their differing edaphic and climatic conditions and point out that these are at least as important as biotic interference in determining their character. Robbins (pers. comm.) now believes the *Ischaemum-Themeda* grasslands are a true edaphic climax.

Another possible and perhaps more likely explanation for the development of such grasslands is that fire from initially small patches of grassland entered and gradually destroyed and pushed back the forest. Under the prevailing conditions of low and seasonal rainfall and physically poor soils the original forest would have been in a precarious balance and hence liable to damage and destruction by fire. The process is rare at the present day, as most ecologically fragile forest has long since been converted to grassland. The occurrence of apparently wind-determined tongues and outliers of forest and grassland on plains and fans, and jagged forest-grassland boundaries on hill slopes both suggest forest destruction directly by fire rather than through gardening.

Seasonality of rain appears to be more important than total annual rainfall. For instance, large areas of rain forest are reported to have been destroyed by fire along the south coast of New Britain around the year 1916 (Lane-Poole 1925: 56), an area with a high annual rainfall, but a marked dry season.

Schoenus-Eriachne sedge-grassland. This community is characteristic of poorly drained coastal and inland plains in the monsoonal part of south-west Papua New Guinea (Plate 23). These plains are very dry in the dry season but are inundated by rain water in the wet season because of impermeable heavy clay soils and lack of surface run-off.

Sedges, mainly *Schoenus* spp., either dominate or form a proportion approximately equal to that of grasses, the most common of which are *Eriachne* spp., *Germainia capitata* and *Ischaemum barbatum*. The dominant sedges and grasses grow in low tussocks,

Plate 23
Sedge-grassland on seasonally inundated flat in south-west Papua New Guinea is bounded by a rim of low *Melaleuca-Banksia-Grevillea* savanna backed by tall mixed savanna. Termite mounds are a characteristic feature in the sedge-grassland.

between which are mainly very low annual grasses, *Selaginella*, and algae. In the dry season the algae form a dry papery layer on the surface.

Herbs, either quick-growing annuals or species able to endure seasonal inundation and recurrent killing of above-ground parts by fire, are many and include tiny species of sundew (*Drosera*), *Utricularia*, and *Eriocaulon*, and a coarse pitcher plant (*Nepenthes* sp.). Widely scattered thin and crooked trees are generally present, growing singly and in small groups, but some areas are treeless. Trees include *Melaleuca cajuputi*, *M. symphyocarpa*, *Banksia dentata*, *Acacia leptocarpa*, *Tristania suaveolens*, *Grevillea glauca*, and *Pandanus*. Shrubs are more common than trees; they are mainly stunted forms of the main tree species and the true shrub *Sinoga lysicephala*.

Termite mounds are a characteristic feature of these grasslands (Plate 23), as also are numerous deer and wallabies, which roam the plains mainly near the coast, where permanent swamps provide drinking water throughout the year.

On the inland plains *Schoenus-Eriachne* sedge-grassland forms an intricate mosaic with savanna and dry evergreen forest. This mosaic pattern is controlled mainly by very slight local variations in height of the terrain. In the mosaic sedge-grassland occupies the worst drained and longest inundated parts, low savanna occurs on slightly higher and better drained sites, and tall savanna and dry evergreen forest cover very gently undulating, relatively well drained terrain. This vegetation pattern, caused by edaphic conditions, has superimposed upon it a general impoverishment of the vegetation due to frequent burning, which prevents recovery towards a true edaphic climax and obscures the natural relationships between vegetation and habitat.

Man-made vegetation

The development of secondary vegetation on abandoned gardens is described in detail in the section on the foothills and low mountains. Early stages of secondary forest are dominated by such genera as *Kleinhovia, Macaranga,* and *Althoffia*. Old secondary forest is commonly characterised by a predominance of one or more of the genera *Cananga, Endospermum, Canarium, Euodia, Laportea, Sterculia,* and *Pimelodendron*. Where bamboo is present in nearby primary forest it is often common in regrowth vegetation. On poorly drained sites late stages of secondary forest may be difficult to distinguish from primary forest, as both tend to have a very irregular structure and an abundance of rattan and lianes. In places low levees and meander scrolls are cleared down to mean river level for growing banana, a crop that can stand short flooding. After such sites have been abandoned they are taken over by swamp grasses.

Plant succession

On the lowland alluvial plains bare land is continually being formed during floods by sedimentation and changes in river courses. The initial colonisation of such land by pioneer plants is followed by successional development, during which ever more new species replace earlier invaders, until finally the mixed forest climax is reached or restored. Three types of succession are distinguished, based on three main different environments: the river bank sere, the river plain sere, and the stream bed sere.

River bank sere. Frequently flooded low levee banks along the lower courses of rivers are colonised mainly by the tall grasses *Phragmites karka* and *Saccharum robustum*. Where the river water is slightly brackish near the coast, *P. karka* and pandans form pure stands. In the fresh water environment further upstream *S. robustum* dominates on the higher, and *P. karka* on the lower sites. One of the first trees to appear on the tops of low banks is *Artocarpus altilis,* the wild breadfruit tree. As levees become higher and are less frequently flooded *A. altilis* is joined by *Octomeles sumatrana,* and there develops a young forest dominated by either or both species. Herbaceous and thin woody climbers abound early in this stage, particularly in open stands. They hang down in large masses from the convenient horizontal branches of *Octomeles* and often completely smother the trees. The ground layer is sparse where flooding is frequent, but elsewhere may consist of dense ferns, gingers, and grasses. Other trees characteristic of forest on alluvial plains, such as *Ficus, Laportea, Nauclea, Kleinhovia,* and *Terminalia,* invade the young forest but remain overtopped by the fast-growing *Octomeles*. The tree density of the young mixed forest is low to begin with, and the lower storeys and shrub layer are open. Low-branched and crooked *Kleinhovia hospita* often dominates below the canopy. In time huge trees of *Octomeles* form an open upper storey up to 60 m high. These trees have large girths, high buttresses, and very wide crowns. There are no seedlings of *Octomeles,* and as the old trees become overmature and die, they are replaced by other species, and the mixed forest climax comes into being.

Plate 24
Pioneering vegetation in inner river curve ranging from swamp grass and mixed herbaceous vegetation near the water's edge and in swale through young *Octomeles*-dominated forest on scroll ridges to mixed forest with *Octomeles* further inland.

Plate 25
Pioneering vegetation revealing and accentuating scroll pattern along Sepik River from the air. Woody vegetation covers highest parts of ridges, and swamp grasses and pandan vegetation grow on low ridges and in swales. Swamp forest, swamp woodland and herbaceous swamp vegetation in relatively stable swamp at bottom.

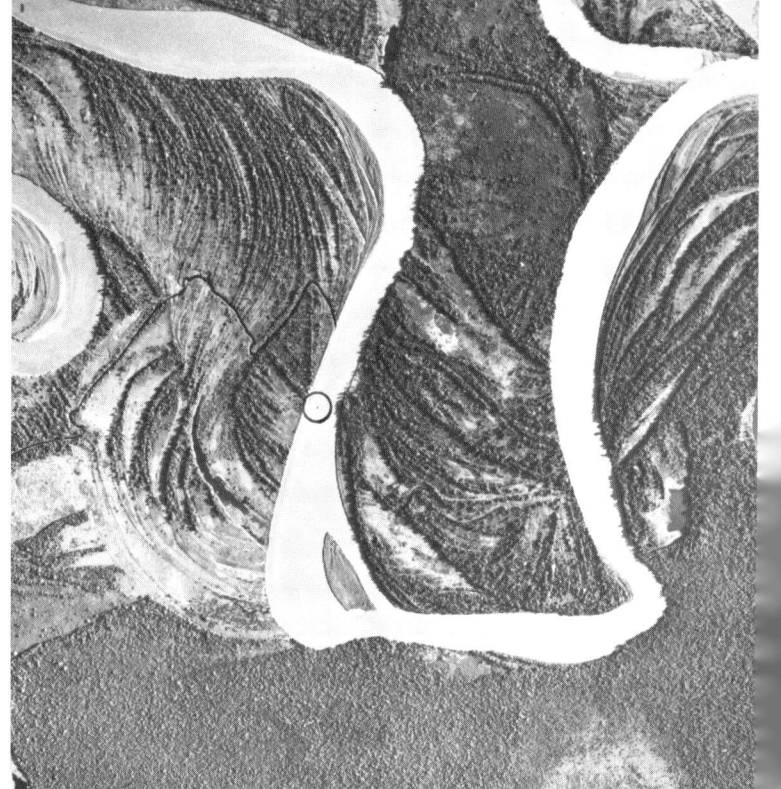

On the inner curves of the larger rivers banks of meander scrolls are built up, consisting of groups of parallel arcuate ridges and intervening swampy swales. The vegetation on these ridges and swales develops in a way similar to that of levees, and is closely related to the topography and age of the scrolls (Plates 24 and 25). *Phragmites* and *Saccharum* colonise the outer rim of the youngest ridge nearest the river channel, and trees of *Timonius, Althoffia, Artocarpus,* and *Octomeles* form narrow, even-aged and increasingly higher stands on successively older ridges. The swales remain unforested for much longer than the ridges. As they become less swampy, pioneering swamp grasses are replaced gradually by Marantaceae and gingers, above which emerge shrubs such as *Leea* and stinging *Laportea,* and trees of *Timonius, Bischofia* and euphorbs.

All stages of the vegetation succession on meander scrolls are often present along a relatively short cross section of a single group of scrolls. In contrast, those on levees can be seen only where the levee banks, stretching far upstream from the river mouth, become gradually higher.

On the island of New Britain many river banks and terraces are colonised by *Eucalyptus deglupta.* This species initially forms pure, even-aged stands on frequently flooded but fast-draining, relatively coarse river deposits. Further development is similar to that of *Octomeles* forest. *E. deglupta* often occurs together with *Octomeles,* is equally fast-growing, and also seeds profusely, but in contrast to *Octomeles* it is not fire-resistant and not suited to conditions of poor drainage. The species grows naturally and easily on various open sites, such as dry river beds, volcanic eruption deposits, and abandoned logging roads.

River plain sere. As drainage conditions on river flood plains improve, the plants first established amongst the initial swamp grasses increase in number and gradually form an almost impenetrable thicket tangled by climbers. Shrub pandans, tall herbs, and sedges usually form a large part of the community, which is otherwise very variable both in the density and in the distribution of its components. Scattered groups of tall rain forest trees grow up on slightly elevated sites, while elsewhere only sporadic, climber-covered, low and thin-stemmed swamp trees are present. As drainage conditions continue to improve, an open type of alluvium forest develops. However, the succession may be deflected locally by changes in the course of the river.

Stream bed sere. Coarse stream deposits in braided upper river channels and outwash fans subject to frequent brief flooding are initially colonised by grasses forming scattered tussocks or growing in small stands. *Saccharum spontaneum, Pennisetum macrostachyum,* and other grasses pioneer on well drained sites, while *Phragmites karka* dominates in swampy situations. The grasses are soon joined by shrubs and trees such as *Ficus,* the exotic *Cassia alata, Albizia falcataria* and *Trema* on well drained sites, and pandans and sago on swampy sites. *Eucalyptus deglupta* is a first pioneer on well drained sites, mainly on New Britain but also in a few localities on mainland Papua New Guinea, for instance south of

Vanimo in the north-west, and north of Table Bay in the south-east.

Casuarina cunninghamiana is a coloniser of bouldery banks and bars that have risen above the level of frequent flooding, and also on sandy, poorly drained to swampy outwash fans. The tree forms dense pure stands, which initially have little or no undergrowth. As litter and sand become trapped between the trees, a shrub and herb layer of grasses, ferns, pandans, and herbaceous creepers develops. When a change in river course moves the stream away from a pioneer stand, the flood waters move more slowly through it. More sand and silt are trapped, and the site becomes more stable. A lower storey of broad-leaved trees establishes itself below the casuarina, which now no longer regenerates, and via a stage of mixed forest with emergent casuarina the forest reaches its mixed broad-leaved climax. However, at any time increased flooding may retard or reverse the succession.

Resources

Some of the best land for oil palm, rubber, and other cash crops in Papua New Guinea is on the alluvial plains and fans. Their extensive grasslands are little used at present but many have a potential for cattle raising. The forests of the plains and fans are favoured by timber exploitation companies because of their easy access, and they yield the bulk of many well known timbers such as New Guinea Walnut (*Dracontomelon puberulum* (= *Dracontomelum mangiferum*)), Taun (*Pometia* spp.), Kwila (*Intsia bijuga*), and New Guinea Rosewood (*Pterocarpus indicus*) (Plate 26). These species are amenable to cultivation in forest plantations, produce seeds regularly and in quantity, and are capable of standing seasonal waterlogging and inundation. Kwila will grow under brackish and fresh water soil conditions, and Rosewood is suited to both everwet and monsoonal climates and, when closely spaced, has a better stem form and greater bole length than in natural mixed forest. All except Kwila are being grown on a trial basis either in pure stands or as natural regeneration released by poisoning unwanted neighbour trees (Papua New Guinea Department of Forests 1973).

Other valuable trees are Erima (*Octomeles sumatrana*) for plywood and Kamarere (*Eucalyptus deglupta*) for timber. Natural stands of both species are rather sparse and of limited extent, but both can be successfully grown in plantations of pure stands. Kamarere, once rare outside New Britain, is becoming widespread through planting. In contrast to the savanna eucalypts it is of Asian origin and has probably reached New Britain via the Philippines.

There is an urgent need for reservation of areas of undisturbed alluvium forest for future study and controlled exploitation on a sustained-yield basis. To quote from a recent address by the Assistant Director, Department of Forests in Papua New Guinea: 'The combination of subsistence farming, intensive forest exploitation, agricultural development, access roading and mining activities, predict that the lowland rain forest as a feature of Papua New Guinea is a passing one. If we act now we may be able to retain some areas as natural museums' (White 1971).

Plate 26
Pterocarpus indicus being felled from scaffold.

Plate 27
Araucaria hunsteinii forest on hills around Musa River basin. The araucarias emerge above a canopy of mixed broad-leaved tree species.

The Foothills and Mountains Below 1000 m

Vegetation and habitat

The foothills and low mountains are an area of broken topography with generally short and often steep slopes and stony ground, a 'medley of wave-like hills', as Lane-Poole (1925) described it looking down from a high mountain ridge.

Forest covers most of the foothills and low mountains, though extensive tracks of grassland and eucalypt savanna are present in low-rainfall areas. However, a high proportion of the forest is secondary almost throughout Papua New Guinea as a result of shifting subsistence farming from both strategic hill ridge and convenient riverside villages. In some areas, for instance the hills and mountains north-west and south-east of Madang, north-east of Kerema, and north of Hood Bay, the landscape is so riddled with patches of garden and various stages of regrowth that very little virgin forest is left.

Grassland covers much of the dry hilly country along the coast in the north-east, and is the main vegetation in many populated intermontane 'rainshadow' valleys, for instance those of the upper Markham, Tauri, and Bulolo Rivers. Eucalypt savanna is most extensive in the hills along the central south coast and is also present inland on the slopes and valleys of the upper Musa River drainage basin and north of the central range near Popondetta.

The forest

Mixed evergreen forest, the main forest type of the hills and low mountains, is not as imposing as that of the plains. Mainly owing to generally less favourable conditions of steep slopes and shallow unstable soils, the forest is less luxuriant, though it shows a greater local variation in structure, species composition and timber volume. Generally, trees are not as large, except the araucarias which stand out by their size and habit (Plate 27).

Mixed evergreen hill forest differs from the average forest on alluvium mainly in structure. The forest canopy is somewhat lower and less variable in height, closure, and crown size. Emergents are also lower and, with the exception of *Araucaria*, rarely reach 50 m. Trees with a very large girth and large buttresses are less common, but the total number of trees with a girth over 30 cm is greater, and there are more trees in the pole and sapling stages. The shrub layer, consisting mainly of slender saplings, has a lower cover, but the herb layer, though very patchy, is generally somewhat denser. Thick woody lianes, rattan, tree and shrub palms, and fleshy climbers and climbing ferns on tree trunks are less common. Tall palms are a normal feature in the canopy and locally emerge above it, but they usually occur in small numbers only. Tree ferns are more common, as also is scrambling bamboo, especially on ridge crests.

Like forest on alluvium, hill forest is very rich in species and very mixed. Most trees present in alluvium forest also occur in hill forest, although in many cases in different proportions. Frequent canopy trees are *Pometia, Canarium, Anisoptera, Cryptocarya, Terminalia,*

Syzygium, Ficus, Celtis, Dysoxylum, and *Buchanania,* but it would not be difficult to extend this list. Some tree genera common in the understoreys are *Garcinia, Syzygium, Diospyros, Myristica, Pimelodendron, Microcos,* and *Gnetum.*

Environmental controls

Because of different environmental conditions there are several variants of mixed evergreen hill forest. The main controlling factors are altitude, topography, climate, and soil and rock type.

Altitude. With increasing altitude the forest canopy becomes more even in height, crown size, and crown spacing. Tree ferns become more common in the shrub and lower tree layers, and herbs such as *Elatostema, Begonia* and, on open sites, prominent orange-pink flowering *Impatiens* feature in the ground layer. Increasing humidity causes the forest to become richer in epiphytes, particularly mosses and ferns. The forest also becomes less rich in species, and trees in the oak family—*Castanopsis* and *Lithocarpus,* the Elaeocarpaceae—*Elaeocarpus* and *Sloanea*—and the laurel family—*Cryptocarya*—become prominent.

Although it is generally assumed that the changes in forest structure and floristics with increasing altitude are mainly due to lower temperatures (see for example Lam 1945: 80), there are indications that cloud cover may be a more important factor. Brass (1964), for instance, observed that throughout Papua New Guinea the lower limit of *Castanopsis/Lithocarpus* forest coincides with the lower edges of the afternoon cloud body, which in the south-easterlies season forms almost daily at remarkably constant and well defined levels on the hill and mountain slopes. In Ecuador, Grubb and Whitmore (1966) studied the changes from lowland rain forest to montane forest, and suggested that these are mainly controlled by the frequency of cloud cover close to the ground. In Papua New Guinea, *Castanopsis* and *Lithocarpus* in many places come down to very near sea level, as along the south coast of New Britain (Jermy 1965), and at low-altitude inland localities in south-west Papua New Guinea (Brass 1938). Both authors attribute this unusually low occurrence to the local conditions of high annual rainfall and almost constant humidity. However, *Castanopsis* and *Lithocarpus* have also been recorded in medium to low rainfall areas in south-west Papua New Guinea (Paijmans 1971).

Topography. Forest on steep and unstable slopes that have thin soils has an open and irregular canopy. Tree dimensions are smaller than average, and many trunks are leaning or bent at the base. Soil erosion causes roots to become exposed, and some trees develop adventitious roots. In contrast, forest on gentle slopes, plateaus, and footslopes has an above average height, tree girth, and crown size and, where drainage and moisture conditions are favourable, approaches lowland alluvium forest in structure. Ridge crests often have a heavier forest than side slopes. Some trees, such as *Araucaria,* tend to be concentrated on crests, while others, such as *Pometia,* are markedly more common on footslopes than on mid and upper slopes. Less obvious floristic variation correlated with topography can be detected by methods of numerical analysis and has been clearly demonstrated in forests in the Solomons, Malaya and Borneo.

A characteristic vegetation consisting mainly of shrubs occurs throughout Papua New Guinea on sandy and rocky banks and beds of streams, sites that are subject to sudden brief flooding by fast-running water. The shrubs have horizontal branches spreading in the direction of the stream, and narrow, willow-like leaves. They are flood-resistant, tough, and firmly anchored by a wide root system. Common genera are *Ficus, Syzygium,* and *Neonauclea,* but many more show the same habit. The type is not restricted to Papua New Guinea, but occurs world-wide. Its members, named rheophytes by van Steenis (1952), also include ferns, grasses, and several more or less permanently inundated aquatic herbs.

Climate. In areas that have an annual rainfall between about 1200 and 1800 mm, and a distinct dry season, the forest has many deciduous and semi-deciduous trees in the canopy. The lower storeys, however, remain evergreen. Such forest is similar in structure to its evergreen counterpart except that the canopy is somewhat more open, and scrambling bamboo is a normal feature. Common deciduous trees are *Garuga floribunda, Intsia bijuga, Terminalia* spp., *Protium macgregorii, Anisoptera polyandra, Pterocarpus indicus,* and *Sterculia* spp. In the shrub layer *Maniltoa, Lunasia amara, Cycas,* and *Desmodium ormocarpoides* with long pendant inflorescences are markedly more frequent than in evergreen hill forest. Scattered grasses, sedges, and ferns form a sparse ground cover. From the air forest rich in deciduous trees commonly can be seen to form a 'fern-leaf' pattern with savanna, the forest occupying lower slopes and valley bottoms, and savanna covering crests and upper slopes (Plate 42).

In areas where the annual rainfall is less than about 1200 mm and the dry season is long and severe, the forest has a low and open canopy dominated by deciduous trees. The tree storey below the canopy consists of deciduous and evergreen trees, and in the undergrowth many shrubs are spiny and scrambling. *Flagellaria* and thin woody lianes, some of which have cork ribs or spines, abound, but epiphytes are scarce. Deciduous trees, in addition to those already mentioned, include *Gyrocarpus americanus, Bombax ceiba, Brachychiton carruthersii, Adenanthera pavonina,* and *Erythrina* sp. Forest of this type is restricted to the mainly limestone hills along the coast near Port Moresby. In places it grades into a woodland or scrub of gnarled trees and shrubs interwoven with lianes. A similar vegetation occurs on coastal limestone in other areas of the South Pacific, notably the New Hebrides (Gillison, pers. comm.).

The severity of the dry season is probably the main factor controlling the frequency of deciduous trees, but altitude and local variations in topography and moisture conditions also play a part. In general, deciduous trees decrease in number with increasing altitude, so that forest near the lower montane zone is virtually evergreen. Deciduous trees may be locally common on ridge crests extending into a zone of otherwise evergreen forest. A comparable situation exists in Malaya where many species of the Burmese-Thai monsoon forest occupy flat lowlands in the seasonally dry extreme north-west of the country, but outside the monsoonal region further south grow on hill slopes and ridges (Wyatt-Smith 1964: 202).

Soil and rock type. Soil appears to have much less influence on the structure and floristic composition of hill forest than altitude, topography, or climate. In Papua New Guinea, however, there is evidence to suggest that forest of below average stem diameter and crown size is correlated with the presence of strongly weathered, acid clay soils, which are fairly common on gently to moderately sloping low foothills (Plate 28). The soil itself does not appear to influence the floristic composition of the forest, but any suspected correlation would be hard to prove because of the complex interaction between soil, topography and climate, and because of other factors such as previous disturbance and chance establishment, which may override and mask any influence by soil.

Plate 28
Dense thin-stemmed forest on low hills north of Lake Murray, Western District. The soil is a deeply weathered strongly acid clay. *Vatica papuana* and *Lithocarpus* are common trees.

Richards (1964) found a close correlation between soil conditions and the floristics and structure of forest types in British Guiana. Richards refers to similar good correlations in other areas and concludes that physical edaphic factors are probably more important than chemical soil properties. In Malaysia various workers disagree about the influence of soils on hill forest. Results of a numerical analysis of Ashton's data from Brunei, Borneo, indicated that on sites of similar topography variation in the species composition of trees with a diameter greater than 10 cm is associated with variation in soils (Austin, Ashton, and Greig-Smith 1972). Wong Yew Kwan and Whitmore (1970) on the other hand report that in an area of forest on gently sloping to flat land there was no indication that either common or rare species are correlated with any of three very contrasting types of soil present. Poore (1968) found that soil differences do not affect the distribution of the commoner and hence probably more tolerant species, but suggests that soil differences may perhaps determine the occurrence of at least some of the rarer species.

Rock type influences the vegetation through the soils and land forms developed on it. Ultrabasic rocks (igneous heavy rocks with low silicon and relatively high magnesium and iron contents) and limestone are of special interest. Throughout Papua New Guinea ultrabasic rocks in particular and, less obviously, limestone appear to be correlated with the presence of dense, thin-stemmed, small-crowned and often low forest. Ultrabasic rocks often have a topography of steep, straight slopes with unstable, shallow and chemically poor soils, and hence have thin-stemmed forest. However, forest tends to be denser and smaller crowned than average hill forest also on ultrabasics that have a gentle topography and deep soils. Shallow soils also predominate on limestone, but are less infertile and penetrate deeper into solution hollows in the rock (Haantjens, pers. comm.).

In Papua New Guinea there is no evidence that any tree species is confined to either ultrabasic rock or limestone. *Casuarina papuana* is commonly predominant in forest on ultrabasic rocks, and occasionally in forest on limestone, but this species has a wide ecological range and is not restricted to any particular rock type or soil. In the Solomons *C. papuana* has been used in air photo mapping as an indicator of ultrabasic rocks (Latter 1960).

In Sabah, Borneo, Fox and Tan Teong Hing (1971) found one tree species of a sample of eighty-one to be typical of forest on ultrabasic rocks. They also stated that until recently the local population have tended to avoid the ultrabasic soils because they are reputed to be toxic.

Evidence from North America (Kruckeberg 1969), New Zealand (Lyon *et al.* 1971), England and Scotland (Proctor and Woodell 1971; Proctor 1971a, 1971b), and many other European countries indicates that ultrabasic areas generally have a stunted and depauperised vegetation, with few but distinctive endemic species, and that in many agricultural areas ultrabasic soils are recognised as infertile and/or toxic. As probable causes are brought forward their unfavourably high magnesium/calcium ratios and the presence in toxic concentrations of the heavy metals nickel, chromium, and cobalt. However, experiments in the laboratory have not been able to resolve the cause of the unusual plant associations on ultrabasics (Lyon *et al.* 1971), and Proctor (1971a) concludes that in the ultrabasic soils examined by him the role of the heavy metals remains enigmatical.

Predominance of certain trees

Many tree species that occur scattered throughout mixed hill forest tend to grow gregariously in certain localities. The most important ones are members of the family Dipterocarpaceae, *Casuarina papuana,* and *Araucaria hunsteinii.*

Dipterocarp forest. The Dipterocarpaceae are represented in Papua New Guinea by the genera *Anisoptera* and *Vatica*, each with one species, and *Hopea* with about five species. This is in marked contrast with Malaya, where some 200 species of dipterocarps are present. The three Papua New Guinea genera are widely but rather patchily spread. Each may dominate over extensive areas, but only rarely, as in parts of south-west Papua New Guinea, do all three together form a large proportion of the canopy.

Pure stands of *Anisoptera polyandra* are mainly stages of young and advanced secondary forest. The species also occurs in primary forest either scattered or more or less gregarious, and grows on both hilly terrain and well drained flats. *Vatica papuana* has a wide range of habitat, growing gregariously on hill slopes as well as in swamps. Various species of *Hopea*, but mainly *H. papuana*, form tall slender forest on hilly terrain. The genus has an irregular distribution throughout the lowlands, and dominates large areas of presumably primary forest in the eastern part of mainland Papua New Guinea. The three dipterocarp species seed and regenerate amply, particularly *H. papuana*, whose seedlings often densely cover the forest floor.

Casuarina forest. Broad-leaved forest with an upper storey of *Casuarina papuana* occurs in low-rainfall localities and on limestone, from almost sea level to well into the lower montane zone. Pure stands of the species form narrow bands between mixed forest and eucalypt savanna. A likely explanation for its gregarious occurrence with and without broad-leaved trees is that *C. papuana*, itself more or less fire-resistant, colonises sites where the original vegetation has been destroyed by fire, and will eventually be

Plate 29
Buttressed trunk of old *Casuarina papuana* tree in mixed hill forest. *Araucaria hunsteinii* tree left of centre.

replaced by mixed forest. The species appears to follow a pattern common to other long-lived pioneers, as it reaches large dimensions (Plate 29) where it is scattered in mixed forest, and does not regenerate under dense shade. Its ecology is probably similar to that of *C. junghuhniana* in Java (van Steenis 1965).

Araucaria forest. The genus *Araucaria* is represented in Papua New Guinea by two species, *A. hunsteinii* (klinki pine), and *A. cunninghamii* (hoop pine). *A. hunsteinii* reaches very large dimensions: a height of over 70 m and a girth of well over 3 m, and towers some 20-30 m above the associated mixed broad-leaved forest (Plate 27). *A. cunninghamii* has a wider altitudinal and geographical range, but usually occurs above *A. hunsteinii,* mainly in the lower montane zone, and does not grow to quite as large a size. Both species are found, often gregariously, on a wide variety of land forms and soils, but tend to be most common on ridge crests and on shallow soils, though this tendency is less marked in klinki pine. The ecology of *Araucaria* in Papua New Guinea is discussed in more detail by Havel (1971), and its distribution has been mapped by Gray (1973).

Some authors contend that the coniferous element in the forests of the South Pacific region is a relic which is very slowly being ousted by mixed broad-leaved forest, the present climax vegetation. This theory is accepted by Womersley (1958) for *Araucaria* in Papua New Guinea, and by Robbins (1962) for podocarps in New Zealand. However, like *Agathis macrophylla* (kauri pine) on Vanikoro, British Solomon Islands Protectorate (Whitmore 1966b), which has a very similar regeneration and growth pattern, *Araucaria* in Papua New Guinea regenerates well, is represented by individuals of all sizes, and appears to be able to maintain itself as a normal and permanent component of mixed broad-leaved forest.

Quantitative data on forest structure and floristics

Quantitative data on tropical rain forest characteristics such as the density, distribution and floristic richness of its trees are rather scanty. For Papua New Guinea the most detailed information is on four sample plots of 0.8 ha each (Paijmans 1970). Three of these plots are in hill forest and one is in a type approaching alluvium forest. In Table 2.2 the Papua New Guinea plots are compared with samples of tropical rain forest in other areas where the same lower tree size limited of 12 inches girth or 10 cm diameter has been used.

Tree density. Table 2.2 shows that the number of trees per ha in Papua New Guinea forest is comparable with that in Borneo and Malaya, and that it is considerably higher than in the plots recorded from Africa and America. The table also shows that tree density tends to be higher in hill forest than in forest on alluvium. In Papua New Guinea this is due mainly to the higher proportion of trees in the lower girth classes in hill forest than in alluvium forest (Fig. 2.3). As the African and American forest plots are mainly on level terrain in the lowlands, their structure is probably similar to that of forest on alluvium in Papua New Guinea, and this may in part explain their lower tree density figures.

Table 2.2 Numbers of individuals and species of trees with a diameter breast height of 10 cm and over in mixed tropical rain forest

Source of data	Plot size ha (approx.)	No. of trees per ha	No. of species on plot*
Papua New Guinea (Paijmans 1970)			
Hill forest	0.8	652	122
	0.8	691	147
	0.8	526	145
Forest on flat river terrace and gentle foot slope	0.8	430	116
Borneo			
Andulau For. Res. (Ashton 1964)			
Ridge	2.0	740	199 (118)
Valley bottom	2.0	640	219 (129)
Sepilok For. Res. (Nicholson 1963)			
Ridge forest	2.0	667	198 (144)
Malaya (Wyatt-Smith 1949)			
Bukit Lagong For. Res. Hill forest	2.0	559	251
Sungai Menyala For. Res. Probably alluvium forest	1.6	489	197
Nigeria			
Okomu For. Res. (Jones 1955)	18.5	451	170
Okomu For. Res. (Richards 1963)	1.5	390	70
Omo For. Res. (Richards 1963)	1.5	521	42
South Cameroons (Richards 1963)			
Southern Bakundu For. Res.	1.5	368	109
Panama Canal Zone (Lang et al. 1971)			
Barro Colorado Isl.	1.5	489	
Brit. Guiana (Davis and Richards 1933-4)			
Moraballi Creek	1.5	432	91
Surinam (Schulz 1960)			
Plot 5, Coesewijne Ri.	1.0		116 (106)
Plot 1, Mapane Cr.	3.0		168 (108)

* Figures in parentheses represent approximate numbers of species on 0.8 ha gauged from available species/area curves.

The size class distribution curve (Fig. 2.3) has the inverted J form typical for mixed tropical forest in general: by far the most trees are in the 12-23 inch girth class; thereafter numbers drop sharply and tend to level out in the high girth classes. In broad terms there is a relation between forest type and the level and slope of its frequency curve. Results from some twenty small plots in Papua New Guinea

Figure 2.3
Number of trees per girth class of one foot, on 0.8 ha
— hill forest
--- alluvium forest

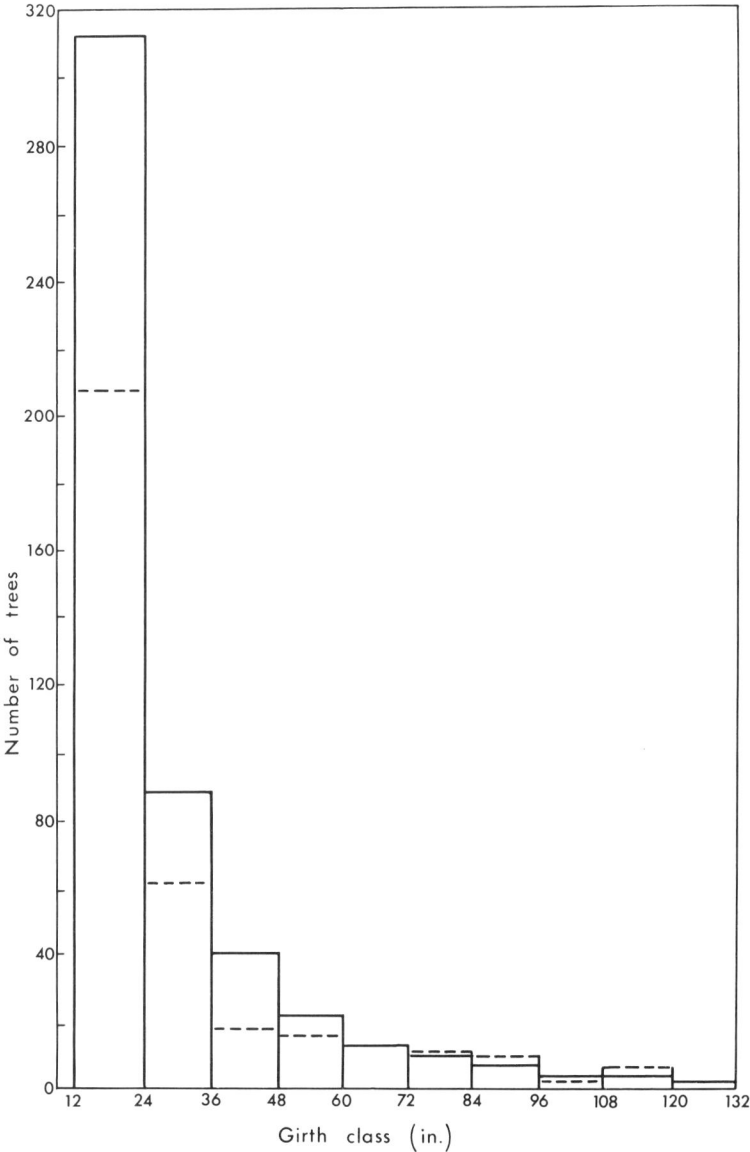

between sea level and 2450 m measured by the author indicate that overall tree density and the number of small trees tend to increase, and the number of large trees tends to decrease, from alluvium forest to hill forest to lower montane forest, but that the variability within each type is very great. With respect to the great variability, this agrees with the results of a much more thorough analysis for Nigerian high forest by Mervart (1972) who found that the frequency curve is a rather ambiguous characteristic of the uneven-aged growing stock, and notes that the frequency distribution is likely to change in the course of time, even if the forest remains undisturbed, and cannot be accepted as a reliable basis for forest typing.

Vertical stratification. To what extent are tree crowns in mixed tropical rain forest arranged in more or less distinct layers? For Papua New Guinea forests Robbins (1969), on the one hand, claims that vertical stratification clearly shows in forest profile diagrams, and furthermore that in its optimum structural development mixed lowland rain forest has three, and only three, tree layers, lower montane forest has only two, and montane forest above about 3000 m only one tree layer. On the other hand, profile diagrams and detailed recording of crown heights in the four Papua New Guinea sample plots did not show conclusive proof of stratification, nor did a study of four lower montane forest sites by Walker (1966).

Richards (1964: 31) accepts that mixed tropical lowland rain forest on a world scale has three tree layers. However, several workers in South American forests found no well marked tree strata (Schulz 1960: 163; Rollet 1968: 532; Vega 1968: 428). Ashton (1965: 146) queries the evidence of stratification in profile diagrams of Malayan forests, and notes that a single diagram represents too small an area to prove the existence of stratification. Fox (1970) on Sierra Leone rain forest notes that the mixed nature of the forest results in continuous variation in heights and proportions, which makes it difficult to describe its structure in terms of strata.

Species diversity. Table 2.2 indicates that the forest on the Papua New Guinea plots is about as rich in species as mixed dipterocarp forest in Borneo and Malaya, and is richer than African and South American forests. A general finding is that species/area curves for

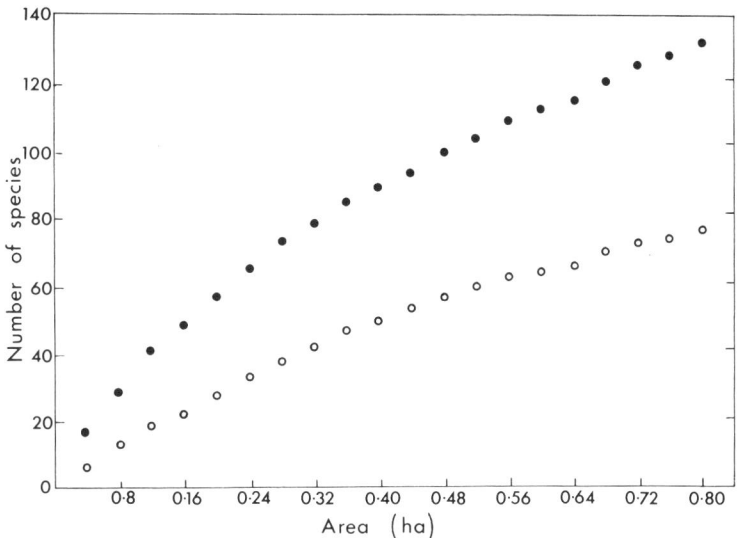

Figure 2.4
Number of species per unit area, average of four Papua New Guinea sample plots of 0.8 ha each
● all trees from 12 in. girth onward
○ all trees from 24 in. onward

mixed tropical rain forest show no clear sign of flattening out when plot sizes are enlarged; in other words, there is little change in the rate at which new species keep coming in (Fig. 2.4). Poore (1964) who sampled 23 ha of hill forest in Malaya reports that the number of species tends to increase merely with the distance from the starting point. This tallies with results from samples elsewhere in Malaya (Wong Yew Kwan and Whitmore 1970) and from the Papua New Guinea plots, which indicate that the number of species common to two plots tends to decrease with the distance between plots. As regards association between species, Jones (1955) in Nigeria and Poore (1968) in Malaya found that within a reasonably uniform area there is no tendency for any pair of the commoner species to grow together or to shun each other's company.

There appears to be a positive correlation between number of species and number of individuals. Generally, for relatively small sample areas, the number of tree species tends to be larger in denser forest. The greater number of species and the initial steeper rise of the curve when smaller trees are included (Fig. 2.4) are in agreement with this tendency. Species-area curves for forest plots in other areas show the same trend, see for instance Schulz 1960: 172 and Lang *et al.* 1971: 396.

A fairly general conclusion is that in tropical rain forest there is little or no evidence of discrete associations of species; instead, the species composition appears to vary continuously (see for instance Schulz 1960; Fox 1970; Walker 1973). There is, on the other hand, a tendency for individuals of one species to aggregate over small areas, in other words to show patchy, or patterned, distribution. This also is reported by various workers in tropical forest, for example Jones (1955), Schulz (1960), Jack (1961), Poore (1968), Fox (1970), and Whitmore (1974), who relate it to small changes of habitat, the history of the forest, chance establishment of seedlings in natural gaps caused by windfall or the death of overmature trees, regeneration near mother trees, or a combination of these factors. In Papua New Guinea chance groupings are probably most often remnants of a late successional stage after disturbance by man.

Apart from small-scale aggregation, there are, in Papua New Guinea, striking and as yet largely unexplained regional floristic differences. Despite apparently similar environmental conditions, some tree genera, for example *Araucaria, Koompassia, Eucalyptopsis,* and the dipterocarps, are very common or even predominant in certain regions, while they are absent in others.

Other vegetation types

The main vegetation types present besides forest are grassland and savanna. Like those of the plains and fans, the grasslands of the hills and low mountains are most extensive in areas that have a marked dry season. The bulk of the grasses is mid-height, but tall grasses are thinly scattered throughout and commonly dominate where soil-moisture conditions are favourable. Many species of mid-height grasses are present, most are of scattered occurrence, but some dominate or codominate locally, and others form more or less pure stands over large areas. The species that most commonly form pure stands are *Themeda australis, Imperata cylindrica,* and *Heteropogon contortus.* Herbs, shrubs, and trees are invariably present.

Eucalypts form by far the most common type of savanna, but other trees usually scattered in eucalypt savanna and grassland in places reach the density of savanna, for instance *Melaleuca* spp., *Albizia procera,* and *Casuarina papuana. Melaleuca* savanna is found, rarely, on hilly terrain on both sides of the central range. *A. procera savanna* is common on Cape Vogel Peninsula, and *C. papuana* savanna occurs on the south-west lower slopes of Mt Suckling.

Plate 30
Fire-disclimax *Themeda australis* grassland on foothills south-east of Port Moresby. Woodland persists in gullies. A patch of eucalypt savanna occurs in middle ground.

Themeda grassland. Tussock-forming *Themeda australis* occurs from sea level to about 2500 m altitude (Henty 1969), but is most prominent in the lowlands (Plate 30). It is the main dominant over large areas of grassland on shallow stony soils in regions that have a low and markedly seasonal rainfall, and shows a definite preference for dry sites, though it is able to endure waterlogging of short duration. *Heteropogon contortus* and *Sehima nervosum* locally codominate with *T. australis*, and other associated grasses include *Sorghum nitidum*, *Arundinella setosa*, *Capillipedium parviflorum*, *Cymbopogon procerus*, *Pseudopogonatherum irritans*, *Eulalia leptostachys* and *Ischaemum* spp. Common tall components are *Ophiuros tongcalingii*, *Saccharum spontaneum*, *Coelorhachis rottboellioides*, *Pennisetum macrostachyum*, and *Apluda mutica*. After burning of *T. australis* grassland the new sward may be temporarily dominated by other grasses, for instance *Capillipedium parviflorum*.

In regions with a higher and less seasonal rainfall *T. australis* is largely replaced by other grasses but maintains dominance on crests and upper slopes, where it is commonly associated with *Arundinella setosa*, though the latter has a wide ecological tolerance and is also found in moist situations.

Scattered herbs occur throughout the grassland. The proportion of legumes is variable, though generally they are rather scarce. Some of the most widespread herbs are *Euphorbia serrulata*, *Buchnera tomentosa*, *Pouzolzia hirta*, *Polygala* spp., and the leguminous genera *Indigofera, Crotalaria, Desmodium, Tephrosia, Atylosia,* and *Pueraria.*

Shrubs and low trees, usually scattered but sometimes in groups, are limited to those that are fire-tolerant. Some of the most frequent are *Melastoma malabathricum, Glochidion, Grewia, Albizia procera, Grevillea papuana, Timonius, Deplanchea tetraphylla, Cycas media,* and, locally, *Casuarina papuana.* A denser woody vegetation known as gallery forest persists along most streams. It is protected from fire by a ground layer of gingers and other tall herbs that do not burn readily.

Imperata grassland. *Imperata cylindrica* is the main dominant in areas that have been recently cultivated. In contrast to *T. australis* it is most common on relatively deep, well drained soils. On such soils and under conditions of regular burning it may maintain dominance; on less favourable sites in monsoonal areas it tends to be gradually ousted by *T. australis. I. cylindrica* has very efficient means of reproduction both vegetatively and by seed, and its underground rhizomes are extremely resistant to destruction by fire and other agents. When subjected to prolonged shading by other plants it thins out and is eventually replaced by shade-tolerant vegetation (Anon. 1944).

Heteropogon grassland. *Heteropogon contortus* (black spear grass, bunch spear grass) is a widespread scattered grass in seasonally dry lowland hill country but also forms pure, usually small, stands. Like *T. australis,* it shows a preference for dry sites and tolerates shallow soils. Both species have seeds of such a shape that they become readily buried in the topsoil where conditions for germination are more favourable than on the surface (Tothill 1969). On dry and stony hills along the north-east coast of mainland Papua New Guinea *H. contortus,* in competition with *T. australis,* commonly dominates on the crests. This is because fires, which often run along crests, do not kill established plants of *H. contortus,* favour the germination of its seeds, and reduce the ground cover of other species (Isbell 1969). A similar ecological pattern for *H. contortus* is reported from Java where the species is regarded as a reliable drought indicator restricted to areas with a distinct dry monsoon (van Steenis 1965).

Eucalypt savanna. Eucalypt savanna occurs from sea level to about 1700 m altitude, but only in areas that have a marked dry season and an annual rainfall below about 1800 mm (Plate 31). The type is mainly found on hilly terrain and is commonly confined to crests and upper slopes, while the lower slopes and valley bottoms remain under forest (Plates 32 and 33). However, in areas that have an annual rainfall of less than about 1300 mm and a severe dry season with up to seven months of less than 100 mm, eucalypt savanna completely covers hills, undulating terrain, and plains.

The top layer consists of one or more of the species *Eucalyptus alba, E. confertiflora, E. papuana,* and *E. tereticornis.* The first three species usually have crooked stems and rarely reach a girth of 1.5 m and a height of 20 m. *E. tereticornis* often has a straight bole and reaches a height of over 30 m. *Nauclea* and *Melaleuca* are two of the rare associated canopy trees. Both tend to be more common on

Plate 31
Eucalypt savanna inland from Pongani on the north coast.

Plate 32
Eucalypt savanna and forest on foothills inland from the south coast, Port Moresby region. The percentage of forest increases with increasing altitude and rainfall further inland.

Plate 33
Airphoto showing areas of grassland reverting to forest. Recurrent fire usually halts or reverses the process.

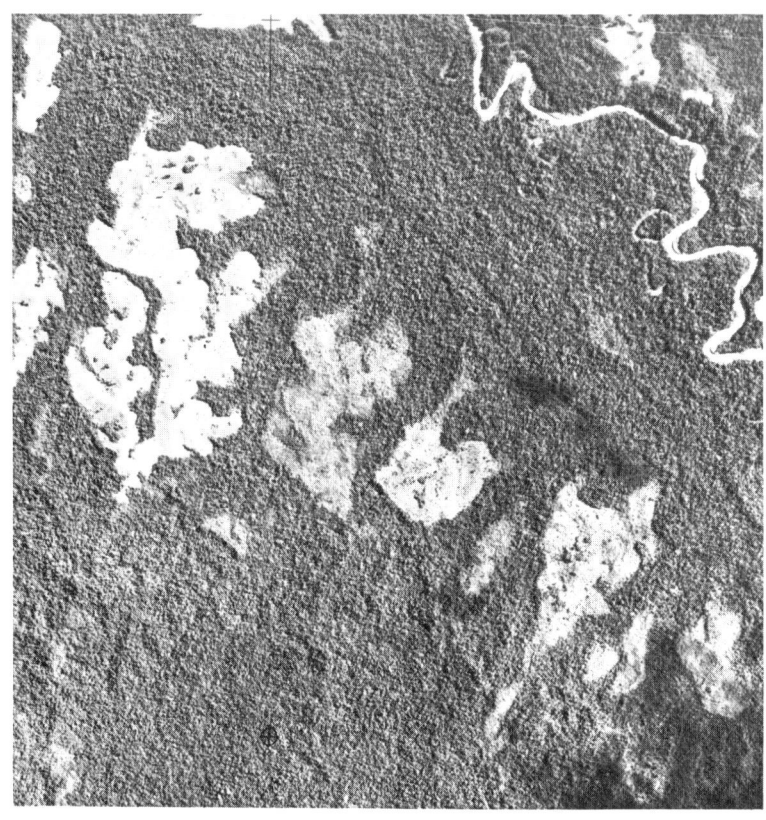

sites with a periodically high ground water level, such as shallow depressions, slopes with a perched ground water table, and seepage areas on footslopes. Lower storey trees and shrubs are usually present, but are thinly scattered. *Albizia procera, Timonius timon, Antidesma ghaesembilla, Desmodium umbellatum,* and *Cycas media* are found mainly at lower altitudes, while *Banksia dentata, Casuarina papuana,* and *Grevillea papuana* are more common in higher parts. Climbers and epiphytes such as *Usnea, Dischidia,* and ferns are rare generally, but tend to be more common at higher altitudes. The grasses and herbs which form the ground layer are those of open grassland in seasonally dry areas, and *Themeda australis* is the main dominant grass.

There appears to be little correlation between species of eucalypt and habitat, and in many places two or three species grow on one site (Heyligers 1966). However, *E. alba* is more frequent on gravelly sites and shallow soils, while *E. tereticornis* tends to dominate where conditions of climate and soil are more favourable. *E. tereticornis* is present over the whole altitudinal range of eucalypt savanna, while the other eucalypts are present only at the lower levels.

The savanna eucalypts are of Australian origin and have spread mainly along the central south coast of mainland Papua New Guinea. They have reached the north coast near Popondetta probably via intermontane valleys of the central mountain range but, rather surprisingly, have not as yet spread to other dry areas on the north side of this range.

Origin and status of eucalypt savanna and lowland grassland

This section deals with eucalypt savanna and those lowland grasslands which have fire as the overriding factor causing their origin, expansion, and maintenance. The ecological status of the Sepik and south-west Papua New Guinea grasslands, which are partly edaphically controlled, is discussed in the section on the alluvial plains and fans.

The view that most grassland and eucalypt savanna of the lowlands is a disclimax vegetation replacing forest and woodland following prolonged gardening and burning is based mainly on the presence of shrubs, trees, and patches of forest within most grassland, and on the proven capacity of grassland and eucalypt savanna to revert to forest when firing is discontinued. By the same token gallery forests are seen as the remnants of a former woody vegetation cover. However, pockets of natural savanna may have existed locally before man began to clear forest on a large scale, on sites unsuited to forest growth, or may have developed after lightning damage in dry forest.

Where for some reason or other regular annual burning is discontinued, the shrubs and trees scattered within grassland increase in density, and genera such as *Alphitonia, Rhus, Commersonia, Decaspermum,* and *Garuga* appear. The grass layer is gradually shaded out and replaced by ferns, gingers, and shrubs, and the grassland reverts to forest via savanna and woodland stages (Plate 34). Small areas of grassland that have escaped burning for some years and show a denser growth of shrubs are not uncommon, but advanced stages of a reversion to forest are much rarer. One such stage is described by Taylor (1964b: 80) from an area that was deserted by the local inhabitants following an eruption of nearby Mt Victory. The conversion from grassland to forest is a slow process, taking considerably longer than that from garden regrowth to forest.

Mature eucalypts are fire-tolerant, but their seedlings and saplings are often badly damaged or killed by grass fires. Normally a few individuals of a group of seedlings survive and thus maintain the community, but savanna is replaced by grassland where fires are very frequent and fierce. Near population centres this process is speeded up by tree-felling and ring-barking. Conversely, eucalypt savanna can revert to forest through a thicket stage of dense lower trees and shrubs of savanna and forest species developing under the eucalypts. A mixed forest with scattered tall emergent eucalypts seen to border savanna at a few localities may be a late stage in the succession to the climax forest without eucalypts.

Boundaries between mixed savanna and forest are commonly poorly defined in south-west Papua New Guinea, but those between eucalypt savanna and forest along the central south coast of the mainland are usually sharp. Eucalypt savanna appears to be a more or less stable community in an environment that may have suffered burning for a longer period than monsoonal south-west Papua New Guinea.

There can be little doubt that the change from forest to savanna and grassland is a degradation of the environment leading to loss of

Plate 34
Pioneering vegetation on lava flow formed during 1914-17 eruption of Pago volcano, New Britain. Ground layer of *Dicranopteris linearis*, groups of *Casuarina papuana*, and single trees of *Eucalyptus deglupta* (left of person) and *Neonauclea* (right of centre).

topsoil through erosion after burning of the grass cover, and loss of nutrients by combustion. However, in Papua New Guinea there is no factual evidence that erosion increases under grassland, nor do grassland soils appear to be generally less fertile than adjacent soils of the same type still under forest. Convincing evidence of soil erosion in savanna grassland is reported for instance by Daubenmire (1972) from Costa Rica.

Man-made vegetation

Abandoned gardens are quickly overgrown by a herbaceous community of garden weeds, grasses, and abundant creepers. This is soon followed and superseded by fast-growing light-demanding, woody plants, particularly some euphorbs, together with tall gingers, ferns, wild bananas and, at higher altitudes, tree ferns. The floristic composition of young regrowth stages varies greatly in detail from place to place. It probably depends mainly on chance establishment and survival of underground vegetative parts and seeds, and to some extent on topography and soil. Provided there is no fire hazard and no regardening, the light-demanding shrubs and trees are gradually replaced by species that are more shade-tolerant, while thin woody climbers succeed the herbaceous creepers. Pandans, palms, and epiphytes initially remain scarce. As more tree species establish themselves, the forest becomes more varied in tree heights and girths. Larger woody climbers, rattan, and other palms appear, epiphytes become well established and more varied, a ground layer of shade-loving herbs develops, and finally the forest becomes similar in appearance to primary forest. However, long after the forest has reached its structural climax, a predominance of certain long-living tree species, some of which entered late in the succession, indicates the secondary nature of the forest. Such trees include *Cryptocarya, Elmerrillia papuana, Endospermum, Euodia, Pimelodendron amboinicum* and *Sterculia.*

During the gardening cycle some trees are left standing or are planted for their edible fruits. These include *Canarium indicum, Terminalia kaernbachii, Dracontomelon puberulum, Pangium edule, Gnetum gnemon, Artocarpus altilis,* the betelnut palm *Areca,* and *Bombax ceiba* for kapok. Other trees and shrubs are planted near villages, such as *Araucaria* because of its attractive shape, *Erythrina* and *Spathodea* because of their showy flowers, and *Cordyline* as a boundary marker. Long after gardens and village sites have been abandoned the unusual frequency of such plants indicates previous human occupation.

Bamboo has already been mentioned as an indicator of human disturbance. It occurs in primary forest, particularly in dry types, and on crests and steep slopes, but tends to spread and become abundant in secondary forest. In places bamboo dominates the shrub and lower tree layers to form the only undergrowth. Various twining, scrambling, and erect species are present. The twining bamboos are thin and may grow up into the canopy, while the scrambling type is thicker and does not grow as high. The erect bamboo is usually thick, commonly grows in stools, and may be either natural or planted.

As the human population of the foothills and mountains largely depends on subsistence agriculture for a living, the practice of shifting cultivation has had and still maintains an enormous impact on the natural vegetation. Villages are widespread throughout much of the zone, and are frequently shifted from site to site. Near the villages the original forest is cleared almost completely and replaced by a patchwork of garden plots and secondary growth. Here the garden rotation cycle is short, and only young regrowth stages are present. Primary forest is cleared for new gardens by people living in temporary shelters often at great distances away from their villages. At these gardens fallow periods are long enough for various stages of advanced secondary forest to develop, and in places the climax forest may become re-established. As shifting cultivation has been practised for centuries, even on very steep slopes (Plate 35), the proportion of secondary forest is higher than is apparent from air photos or at first sight on the ground. On average it may well be over a quarter of the forested area.

Plant succession

Much natural disturbance in hill forest results from landslides and slumps which are very common in Papua New Guinea and are caused by earth tremors or heavy rain or both. Landslides take a long time to become revegetated when all the soil has slipped down, but are quickly overgrown by grasses, garden weeds, tree ferns, and regrowth shrubs where some soil is left. Forest developed on

Plate 35
Interior of lower montane forest at 2700 m

unstable slopes subject to frequent earth movement remains irregular, open, and rich in rattan.

Volcanic activity is also an important cause of natural disturbance, both on volcanic cones and on blast areas and lava flows. On the cone of an active volcano the vegetation is determined mainly by the edaphic conditions and the availability of moisture. The upper part of a cone is generally bare of vegetation because of lack of moisture, hot and crusted ground, and the presence of numerous steam vents and solfataras. Lower down the first tufts of club moss (*Lycopodium cernuum*) and the fern *Dicranopteris linearis* appear in shallow gullies, where the surface crust has collapsed. Further downslope they are joined by scattered trees of *Casuarina*, *Trema*, and *Timonius*. Their bases are surrounded by patches of markedly taller and denser ferns, indicating a more favourable microenvironment for plant growth underneath the trees. Grasses such as *Saccharum spontaneum* and *Imperata cylindrica* may dominate on sites where finer, moisture-retaining material has been deposited (van Royen 1963b; Paijmans 1973a).

On blast areas away from volcanic cones there is usually a variety of seral communities ranging from sparse ferns and club mosses, through a variable cover of grasses, to scrub, woodland, and forest (Taylor 1957). The structure and composition of these communities depend on the nature and thickness of the volcanic deposits, and the extent to which they have been eroded and redistributed. The floristic composition depends largely on opportunity of establishment and differs on different volcanoes. Some trees that are able to colonise soilless or badly eroded areas are very widespread, for example *Trema, Timonius, Casuarina* and *Pipturus;* others such as *Octomeles sumatrana, Albizia falcataria, Ficus* spp., and *Eucalyptus deglupta* are common mainly on ash and pumiceous debris fields (Taylor 1957; Paijmans 1973a).

On blocky lava flows ferns and woody and herbaceous plants begin growing more or less simultaneously, and in an apparently rather haphazard fashion, from crevices where moisture and finer material collect. Thus, bare areas alternate with fern communities of *Dicranopteris linearis* and *Nephrolepis* spp., and stands of *Casuarina papuana, C. cunninghamiana, Timonius, Pipturus,* or *Rhus taitensis* (Plate 36). In woody stands vines of *Hoya* spp. abound, ferns and locally grasses form most of the undergrowth,

Plate 36
Elfin woodland on summit of Mt Oga, 2590 m

and ground orchids are common. Initially both woody and herbaceous stands are often monospecific, but as the environment becomes habitable for other species a mixed forest develops. The succession is likely to be most rapid starting from woody pioneer vegetation, as fern thickets in particular tend to hinder the establishment of other plants (Paijmans 1973a).

On Bougainville Island a continuous scrub of dense climbing bamboos and emergent tree ferns of the genus *Cyathea* covers large areas of steep ash-mantled ridges from almost sea level up to about 1500 m. Pioneer trees are scattered through the bamboos and tree ferns, an open herb layer of ferns and grasses is present, and the ground is covered with a thick springy layer of old leaves and stems. Most of this scrub is probably a successional stage following destructive volcanic eruptions, but locally it may be an edaphic climax. The dense growth of bamboo prevents or retards further development to forest (Heyligers 1967).

Resources

The largest forest industry complex, based mainly on *Araucaria*, is located at Bulolo (Papua New Guinea Department of Forests 1973), at an altitude of 700 m and connected by road to the port of Lae. Many small mills operate on a basis of selective logging, and supply sawn timber for an expanding local market.

Until recently large-scale timber exploitation has been uneconomic in most of the foothill and low mountain zone because of access problems. Owing to modern technological advances, tropical hardwood mixtures can now be utilised for pulping and paper making, hence a much wider area of forest than before has come within reach of profitable exploitation by a combined timber and wood chip industry. In view of the great world demand for paper and the need for cash in developing Papua New Guinea, the near future is likely to see a rapidly expanding exploitation of the country's forest resources (Endacott 1971). Large clear-felled areas will have to be reforested each year. There will also be a pressing need for areas of relatively undisturbed hill forest to be set aside for conservation or to be brought under sustained forest management, preferably including relatively scarce types such as mixed *Araucaria* forest.

A number of useful trees are particularly common in secondary forest. Some, such as *Elmerrillia*, *Anisoptera* and *Pometia*, produce excellent timber; others, such as *Albizia*, *Endospermum* and *Sterculia*, are fast-growing softwood species suitable for crates and matchboxes. Such trees are likely to offer good prospects for cultivation in pure stands because of their natural clustered occurrence in secondary forest, and their fast growth rate.

Extensive plantations of *Araucaria* already exist near Bulolo. To overcome the serious problem of insects and other pests in these monocultures it may be necessary to grow the tree under more natural conditions in mixed *Araucaria* broad-leaved plantations.

The grasslands have become an integral and useful part of the New Guinea landscape and should become increasingly productive when their grazing potential is developed under proper management (Gillison 1972).

The Lower Montane Zone

Vegetation and habitat

The lower montane zone, between 1000 and 3000 m, is characterised by long slopes, spurs and ridges leading up to the great divide. The hot, oppressive climate of the lowlands has given way to a cooler atmosphere more agreeable to the European traveller, though persistent low-lying mist and drizzling rain often spoil his pleasure.

Forest covers most of the area. Lower montane forest is moist, and at the higher levels almost constantly dripping wet. It is also quiet and relatively lifeless. The noisy hornbills and white cockatoos of the lowlands are no longer heard, and no mosquitoes or ants worry the traveller. Many stumps of dead trees remain standing, and a thick springy layer of mosses, humus, leaves, surface roots, and fallen branches covers the ground. Access through the forest is often hindered by fallen logs and thick branches, unexpected hollows under surface roots covered by moss, a dense ground layer, and, last but not least, scrambling bamboo which by its hard siliceous yet flexible stems defies any but the sharpest bush knife. The visibility through the forest is more limited than in most lowland forest types.

Vast stretches of grassland are present particularly in the higher regions. Areas of gardens mixed with grassland cover hundreds of square km in the densely populated Highland Districts of Papua New Guinea, where patches of forest remain only on ridges above 2700 m, the upper limit of native agriculture. A very small proportion of the total area is covered by swamp vegetation in intramontane basins.

The forest

Mixed lower montane forest is evergreen and has a canopy between 20 and 30 m high (Plate 37). It is smaller crowned and more even in height than average lowland hill forest, and is also more densely closed and more regular. Leaves are mainly simple, often leathery, shiny, and dark green, and have entire or serrated margins. The average leaf size is smaller than in lowland hill forest. Tree density is

Plate 37
Interior of *Nothofagus* forest at 2000 m

often very high, but the average girth is smaller than in forest at lower altitudes. Many trunks are low-branched, and bent or leaning, and old trees have thick, crooked, and often dead branches.

Frequent canopy trees belong to the families Fagaceae, Lauraceae, Cunoniaceae, Elaeocarpaceae, and Myrtaceae, the genera *Ilex, Dryadodaphne,* and *Planchonella*, and, particularly in the upper levels, conifers. In the lower tree storeys are found *Garcinia, Astronia, Polyosma, Symplocos, Sericolea, Drimys, Prunus, Pittosporum,* and Araliaceae. Notably absent from the higher regions are Meliaceae, Burseraceae, Annonaceae, Leguminoseae, Dipterocarpaceae, and Ebenaceae, families that are well represented in the lowlands.

The shrub layer is generally denser than in average lowland hill forest, but the density varies greatly with its composition and with the height and density of the tree layers. In many places it consists of a great number of slender saplings, and in other places tall ferns and gingers are more common than woody undergrowth. Various species of Rubiaceae, Myrsinaceae, Melastomataceae, *Eurya, Cyrtandra, Saurauia,* and *Piper* are nearly always present.

The ground flora, mainly consisting of mosses, ferns, herbs, lycopods, and seedlings, is also variable in density. In places mosses almost completely cover the ground and the fallen logs and branches, in other places ferns or *Elatostema* form a dense layer. The tall-stemmed moss *Dawsonia*, shaped like a miniature pine tree, is conspicuous, and near forest borders and in glades *Sphagnum* moss cushions may be found.

Lianes are less common than in hill forest. Thin woody ones abound in places, but thick woody climbers are seldom met. Climbing rattan and other palms are rare, and above about 2200 m are absent. However, a characteristic thin-stemmed climbing bamboo of the genus *Nastus* in many places forms dense tangles, and a scrambling *Rubus* often covers the ground and shrubs near forest edges. Other climbers include the pandan *Freycinetia*, Gesneriaceae, *Lycopodium* and ferns. Epiphytic and ground mosses become abundant with increasing altitude, as do tree ferns and epiphytic ferns. Orchids are usually quite common in tree crowns, low on the trunks, and on the ground. Stilt roots and adventitious roots are found in places, but heavy buttresses are all but absent.

Stilt-rooted pandans, often very tall and reaching into the canopy, occur both singly and in groups. Some of these are spared during forest clearing and are also planted in gardens, as their oily seeds with carotene-rich outer layers are highly valued as food.

Environmental controls

The main factors influencing lower montane forest are climate and altitude, topography, and, to a lesser extent, soil and rock type.
Climate and altitude. The frequency and duration of low cloud cover in particular appear to have a marked if not overriding effect on the physiognomy of lower montane forest. In regions frequently covered by low cloud, the forest begins to present a lower montane appearance well below the normal lower limit for the type, whereas

above a strongly monsoonal lowland zone lower montane features begin to show at an altitude much higher than normal. Lane-Poole (1925: 35) remarks that 'in the present state of our meteorological knowledge it is not possible to set out the limiting (climatic) factors for each belt' and this applies almost as much at the present time as it did fifty years ago. Although climatic variations cause mainly physiognomic changes, there is some evidence that they also affect the floristic composition of the forest. Brass (1941b: 289) gives an illustration and vivid description of a 'mossy forest' on a ridge at 900 m altitude characterised by *Nothofagus, Phyllocladus* and *Astronia,* genera that are usually common only at higher altitudes. In general, the higher the altitude of the forest, the longer it is enveloped in cloud and the more pronounced are characteristic lower montane features, especially a general mossiness. This has led to such forest being known as 'cloud forest' and 'mossy forest' or 'moss forest'.

The floristic composition of the forest also changes with increasing altitude. The number of individuals in certain taxa increases, and in the higher parts of the zone the total number of species present decreases. Lower in the zone the oak family, with *Castanopsis* and less commonly *Lithocarpus,* often predominates; *Nothofagus,* or southern beech, is prominent mainly between 1500 and 2600 m; and in the upper part conifers, Myrtaceae, and Elaeocarpaceae become increasingly common. This pattern of changing predominance shows clearly from the air by yellowish brown colours for *Castanopsis,* grey tones and patches of dead trees for *Nothofagus*-dominated forest, and increasingly darker tones at the highest altitudes indicating an abundance of conifers. The forest in the highest part of the zone is also characterised by the families Myrsinaceae and Ericaceae and, especially along the margins, by genera such as *Drimys, Carpodetus, Olearia* and *Schuurmansia.* Near forest borders with grassland in these high areas many tree crowns are adorned with woody parasites belonging to the families Loranthaceae and Santalaceae.

The altitudinal zonation and the changes in the structure and composition of the vegetation in Papua New Guinea are very similar to those in other tropical high-mountain areas (cf. Troll 1957). The floristics are strikingly similar to the Andes of South America, where at comparable altitudes many of the genera present are those that are characteristic of montane forests in Papua New Guinea (Beziat 1968; Mann 1968).

Forest plants at high altitudes are characteristically more flexible in habit than in the lowlands, where forest dwellers tend to be more strictly bound to one life form. In the higher regions of lower montane forest it is quite common for individuals of one species to occur in both tree and shrub form, like several species in the Ericaceae, and in cases also as a semi-climber or even liane. There is also a greater versatility within the genus. An example is *Pittosporum,* which at high altitudes is represented as a tree, shrub, thick woody liane, and epiphyte. Orchids that are normally epiphytic are also present in the ground flora, where they may have germinated or continued to flourish after falling off a tree. In the lowlands fallen epiphytic orchids do not appear to survive for long on the forest floor.

Topography. On the gentle slopes of plateau areas and mountain saddles at altitudes over 2000 m many forest trees grow to a height of 40 m and have girths over 1.5 m. In contrast, on ridge crests the forest canopy may be 20 m or less high even at altitudes below 1500 m. Narrow ridge crests have steep slopes, shallow soils, are more exposed to wind, and are shrouded in mist more often and for longer periods than the sides of the ridges. Such crestal areas commonly bear a forest of closely-spaced, thin-stemmed, dwarfed, crooked and gnarled trees grading into scrub towards the upper limit of the zone and known as 'elfin woodland' (see for example Troll 1957: 43; Grubb et al. 1963:597) (Plate 38). As in the lowlands, forest on steep slopes tends to be irregular and open, with many trees leaning or bent at the base, and climbing bamboo usually abundant.

Plate 38
Sequence of natural(?) grassland on flat valley floor, fire-disclimax tree fern savanna on lower slopes, subclimax(?) coniferous forest on higher slopes, and mixed lower montane forest on upper slopes and crest. In places a narrow shrub zone forms the transition between savanna and forest. Pyramidal crowns belong to *Papuacedrus papuana*. Alt. ca. 2600 m.

Soil and rock type. The influence of soil, or the lack of it, is largely correlated with topography. As in the forests of the lowlands the underlying rock type appears to have only a minor influence, except where ultrabasics and limestone are present. On these rock types the forest tends to be denser and smaller-crowned, and consequently to have a lower average girth than on other rock types. Air photos and scarce ground observations indicate that conifers and *Casuarina papuana* tend to be more frequent on ultrabasics, but as in the lowlands, no species are known to be restricted to this rock type.

Predominance of certain trees

The main trees that tend to form pure stands in lower montane forest are *Castanopsis acuminatissima, Nothofagus* spp., and various conifers.

Castanopsis forest. *Castanopsis acuminatissima* forms almost pure stands on ridge crests and upper slopes between about 500 and 2300 m altitude. Forest dominated by *Castanopsis* has a dense and even canopy, an open shrub layer, a very sparse ground layer of herbs, and a thick carpet of fallen leaves on the forest floor. *Castanopsis* coppices freely from the base of the trunk, thus many trees have a ring of coppice shoots around the main trunk, or branch out into several stems at or near ground level. Rather surprisingly, the species is occasionally found scattered in swamp forest.

Nothofagus forest (Plate 39). About nineteen species of the genus *Nothofagus* are present in New Guinea. The genus is known to occur between 600 and 3100 m a.s.l., but assumes dominance mainly between 1500 and 3000 m (van Steenis 1971). Outside New Guinea *Nothofagus* has been collected at 100 m altitude on New Caledonia, and it is likely that also in New Guinea its range will be found to be wider as more collections are made.

Like *Araucaria, Nothofagus* has a very patchy distribution pattern, but tends to be most frequent on ridge crests and upper slopes. Though a common component of mixed lower montane forest, it may be frequent in one locality but absent from an adjacent one of apparently similar habitat conditions. It occurs in isolated, often sharply delimited groves on side slopes, emerging 10-20 m

Plate 39
Tree fern savanna of *Cyathea* at ca. 2600 m

above the surrounding mixed forest, but may also form an almost continuous cover of hundreds of square km as in the Southern Highlands and on a plateau in central New Britain. It is also found gregariously on limestone pinnacles and doline rims, and occasionally in swamp forest.

In most areas where *Nothofagus* occurs, scattered young trees of all sizes are present. Regeneration takes place in open forest as well as under shade, but is densest and most vigorous in small openings. *Nothofagus* is reported to regenerate prolifically after disturbance of the forest in Irian Jaya (Brass 1964: 184). It has ectotrophic mycorrhiza, a common feature of Fagaceae, and this probably helps it in its pioneering habit (van Steenis 1971: 72).

Several authors have noted that regeneration is found only under or just beyond the crowns of mature trees (Walker 1966; Kalkman and Vink 1970; van Steenis 1971). It is perhaps not commonly known that much regeneration of *Nothofagus* consists of root suckers, which may explain its limited and slow spread. However, seedlings are also usually present. Whether or not percentages of seedlings and root suckers differ for different species of *Nothofagus* is not known.

Walker (1966: 520) notes that in the New Guinea Highlands *Nothofagus* patches 'superimposed' on other vegetation types appear to be relic rather than pioneer. Mann (1968: 176) holds a similar view with respect to the *Nothofagus* forests of the South Chilean Andes, which he considers to form relict 'islands' as yet undisturbed by more modern and aggressive communities. The presence of extensive pure stands of mature *Nothofagus* in Papua New Guinea remains enigmatic, and the patchy distribution of *Nothofagus* will not be fully explained until more is known about the ecology of the genus.

A common feature of mature *Nothofagus* stands are patches of dead and dying trees, for which no obvious cause has been found. It

Plate 40
Airphoto showing grassland (light-toned), tree fern savanna (grey-toned) and lower montane forest (dark-toned). Altitude ca. 2800 m.

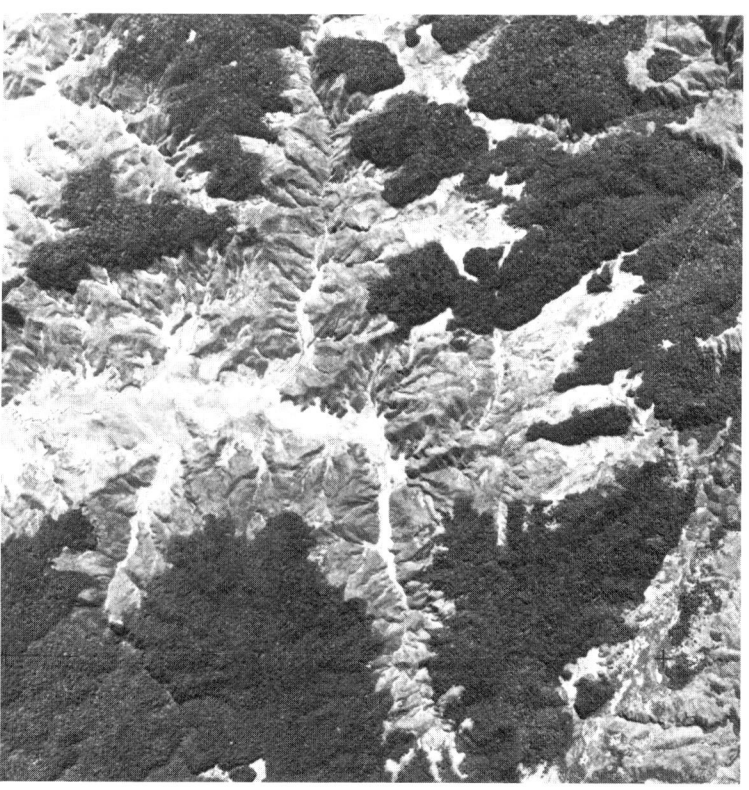

has been suggested (Robbins and Pullen 1965: 105) that groups of even-aged trees die off together on reaching overmaturity. Another possible cause is that a group of trees grown from root suckers of one mother tree die together after infection by a root fungus.

Coniferous forest (Plate 40). In many places above about 2400 m conifers of the genera *Podocarpus, Dacrycarpus, Papuacedrus, Phyllocladus* and *Araucaria* dominate in the canopy and emergent tree layers. Although generally smaller crowned than their broad-leaved associates, many conifers reach a girth of well over 1.5 m, even at altitudes above 3000 m, and have a better stem form. Emergent trees of *Papuacedrus papuana* and *Araucaria cunninghamii* (hoop pine) are easily recognised from afar, the former by its narrow pyramidal crown, and the latter by its rather open crown with horizontal branches hung with grey streaks of 'beard moss' (the lichen *Usnea*). *P. papuana* has a wide ecological tolerance and is able to regenerate under dense forest, in the open, and on steep stony slopes. Many trees are killed by bark stripping, as its bark is especially favoured for roofing.

Belts of almost pure coniferous forest locally form transition zones on mountain side slopes between grassland below and mixed forest above (Plate 40), and also occur on upper slopes of large dolines in limestone country. In such situations the hardy conifers may be pioneering where broad-leaved forest is absent owing either to fire or to frost.

Other vegetation types

Grassland covers the next largest area after forest. Other, less important vegetation types in the lower montane zone are tree fern savanna, sedge-grass swamp, swamp forest, and palm forest. Up to about 2000 m mid-height lower montane grasslands occur mainly in populated intermontane valleys, and here are dominated by various combinations of the same grasses that are present in the foothills and low mountains environment. Between about 2000 and 2500 m grasslands are relatively small and scattered; they may be dominated by species such as *Eulalia leptostachys, Ischaemum* spp., *Arthraxon ciliaris*, and *Imperata*, and usually have a scattering of grasses that become prominent only at higher altitudes. Places where grasslands below 2000 m merge with those above 2500 m, such as in the mountains south of the Markham and Ramu Rivers, are uncommon. Above about 2500 m grasslands become progressively more extensive with increasing altitude, at the expense of forest; they are floristically and structurally very different from those of the hills and low mountains.

Mid-height lower montane grassland above 2500 m. This grassland is generally little over 0.5 m high, and compared with lowland grassland is more tussocky and richer in herbs, though legumes are absent. The major dominant grasses are *Danthonia archboldii* and *Deschampsia klossii*. Other grasses, such as *Agrostis reinwardtii*, other species of *Danthonia, Dichelachne novoguineensis, Deyeuxia* spp., *Anthoxanthum angustum*, and *Arundinella furva*, locally dominate or codominate, and the pattern differs strongly from place to place. *Danthonia archboldii* shows a preference for well drained sites, while *Deschampsia klossii* tends to be more prominent on wet terrain. Both species form tussocks with large crowns which under favourable conditions are up to 1 m high and wide. Where soils are shallow or poorly drained, the tussocks are much smaller and lower. Sedges are common, and include a tall *Gahnia* which forms scattered tussocks among the grasses, and *Machaerina rubiginosa,* which on wet sites may grow in pure stands. Herbs are most abundant on shallow and wet sites. *Gentiana*, *Potentilla*, *Ranunculus*, *Anaphalis* and the fern *Gleichenia* are among the many herbs that are nearly always present. Trees ferns are most prominent near forest edges, and scattered low shrubs occur throughout, but are commonest on relatively well drained sites. Very frequent are the shrubs *Styphelia suaveolens* and *Hypericum macgregorii,* both of which locally grow dense enough to form a heath that replaces the grasses.

Miscanthus grassland. Large tracts of abandoned garden land mainly between 1500 and 2500 m are covered by tall grassland dominated by *Miscanthus floridulus* (sword grass). *M. floridulus*, a cane grass with sharp finely serrated leaves, has spread widely through shifting cultivation, especially in the Western and Southern Highlands, growing either in pure stands or in mixture with *Imperata cylindrica.* It forms a dense cover usually about 3 m but up to 5 m high. Small natural stands grow along swamp margins, on river banks, and in forest glades.

Tree fern savanna. A type of savanna consisting of cycad-like tree ferns of the genus *Cyathea* and grassland covers large areas mainly between 2700 and 3300 m on many high mountains. The type is commonly present between forest on upper slopes and swampy grassland on footslopes and valley floors. On valley floors tree ferns are either absent or are restricted to stream banks (Plates 40-42).

The Cyatheaceae are essentially forest dwellers, indicative of moist tropical mountain forest (Troll 1959). However, in Papua New Guinea some members of the genus *Cyathea* thrive in open grasslands where they are subject to periodic burning, occasional drought, and great daily temperature changes. Their thick fibrous 'bark' and scale-protected frond buds make them to a large extent both fire-tolerant and frost-hardy. Tree fern savanna is not mentioned in the vegetation classification systems of Burtt Davy (1938) and Fosberg (1961), and is probably unique to New Guinea (Paijmans and Löffler 1972).

Sedge-grass swamp. Communities dominated by sedges and grasses occur above about 1800 m in swamps occupying intermontane basins, local depressions in valley floors, and seepage slopes, where either standing or slowly moving water is permanently at or just above the surface. The sedges and grasses are usually low, but in places are over one metre high. They are interspersed with various swamp herbs, and scattered dwarf shrubs are locally present on little hummocks. Many different sedges are present, especially in stagnant swamps, and they commonly make up most of the ground cover. One of the most common sedges is *Machaerina rubiginosa,*

Plate 41
Satellite photo showing lower and upper montane forest on Mt Sugarloaf, ca. 3960 m (centre), Mt Giluwe, 4368 m (below centre), Mt Hagen, ca. 3800 m (right of centre). Grassland covers summit areas and most of the land below the upper limit of agriculture. Lake Kutubu at 800 m in bottom left corner.

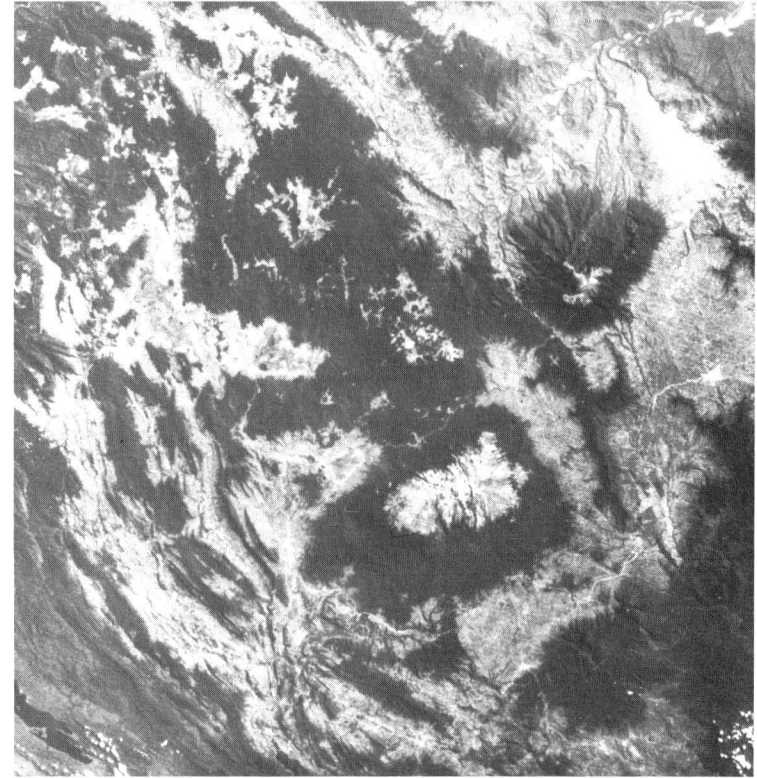

Plate 42
Airphoto showing 'fern-leaf' pattern of eucalypt savanna and forest. This pattern covers extensive low hilly tracts inland from the south-central coast. The proportion of forest increases in the higher hills towards the north-east (top right).

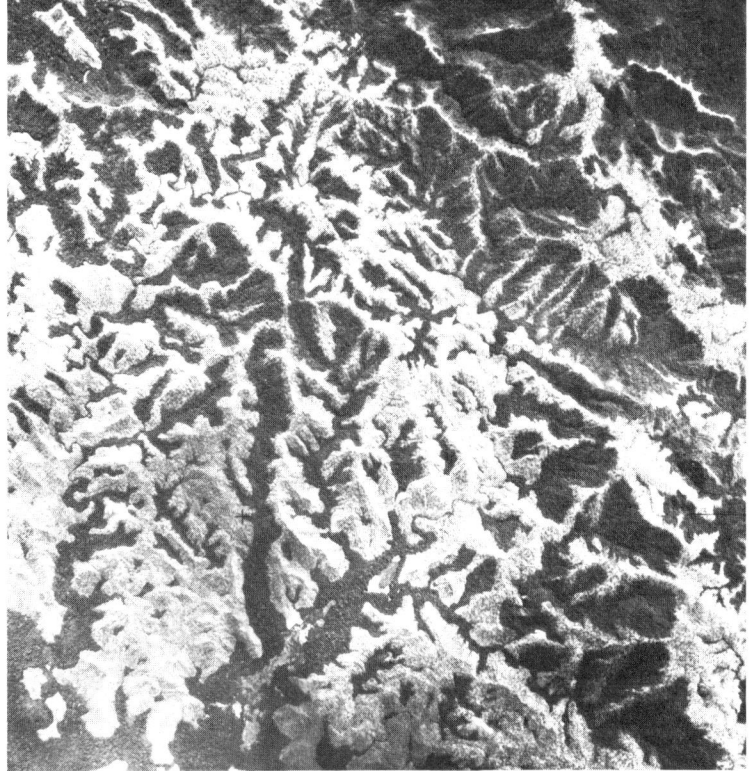

which occurs mainly between 1800 and 3000 m, often in pure stands. Characteristic grasses are *Arundinella furva* and species of *Isachne* and *Dimeria,* mainly between 1800 and 3000 m, and *Agrostis reinwardtii* from about 1800 to well over 3000 m.

Many of the grasses in high-altitude swamps have a wide ecological range; they tolerate lower montane swampy conditions, but find their optimum elsewhere. *Leersia hexandra,* for instance, is dwarfed near its altitudinal limit, at about 2300 m, but grows to a height of 0.5 m or more in lowland swamps. Other grasses, such as *Deschampsia klossii* and *Anthoxanthum angustum,* are stunted on swampy sites, but grow to over 1 m and 75 cm tall respectively on relatively well drained slopes. Even more marked is *Miscanthus floridulus* which grows to 5 m on well drained sites but can reach maturity at only 25 cm in bogs above 2500 m.

Phragmites grass swamp. *Phragmites karka* commonly forms pure stands in seepage areas on slopes and on flat valley floors to over 2500 m a.s.l. The stands are small on slopes, where seepages are very restricted in area, but can be extensive on valley floors. *P. karka* also occurs associated with *Miscanthus floridulus* along river banks and swamp margins, and in very shallow swamps. Both grasses usually form large hummocks rising well above water level.

Swamp forest. Lower montane swamp forest grows in small patches and bands fringing swampy intermontane basins occupied by grass or sedge swamp. The forest usually has a low and open canopy over a dense layer of small trees and shrubs, and a sparse herbaceous ground cover. Most trees grow on hummocks separated by deep pools of water. Common trees include *Syzygium* and other Myrtaceae, *Garcinia,* conifers, and locally *Nothofagus perryi.* Some stands are dominated entirely by conifers, particularly the genus *Podocarpus.*

Palm forest. On Bougainville Island large areas of steep terrain above 1200 m are covered by an irregular and rather open forest about 20 m high, dominated by palms, mainly *Gulubia* sp., in the canopy, and by pandans in the lower storeys. On very steep slopes the pandans in the lower storey are largely replaced by bamboo. Tree ferns and other ferns feature in the shrub layer, and ground mosses, epiphytic mosses, and other epiphytes abound. The habitat is characterised by high humidity, and high precipitation, and is shrouded in cloud during part of almost every day (Heyligers 1967). Tall tree palms of the genus *Gulubia* are fairly common throughout the lowland and lower montane rain forests of Papua New Guinea, but outside Bougainville are usually of scattered occurrence.

Origin and status of the high-mountain grasslands and tree fern savanna

Most high-mountain grassland below the tree line is probably secondary after forest. Possible causes for the destruction of high-mountain forest are clearing by man, frost, and fire. Clearing of forest and felling of individual trees are not significant above 2700 m, the upper limit of native agriculture, and cause only minor damage. Frost damages forest edges and probably prevents forest regeneration once forest has been replaced by grassland. According to van Steenis (1962b) frost does not affect tall closed forest and is not a primary cause for its disappearance. In Papua New Guinea, however, there is some evidence to the contrary. Brown and Powell (1974) report extensive damage to swamp forest and beech forest by a series of severe frosts during 1972. In the same year areas of mixed lower montane forest were killed adjacent to roads, and in deep narrow gullies and similar situations where cold air settled in depth (K.J. White, pers. comm.; J.S. Womersley, pers. comm.). Frost pockets, i.e. low-lying treeless areas where cold air drainage prevents forest growth, are probably of only local importance.

There are, on the other hand, many well documented observations of fire damaging and destroying high-mountain forest (Lane-Poole 1925; Archbold and Rand 1935; Brass 1941b; Wade and McVean 1969; Kalkman and Vink 1970; Paijmans and Löffler 1972). Although the vegetation above about 2500 m is wet for most of the time, exceptional dry periods occur at times, probably at intervals of many years. During these periods grassland, swamp vegetation, and even forest become dry and flammable, and grass fires lit by hunters and travellers become extensive and spread into forest. Initially, swampy patches of natural grassland have probably been the centres from which grass fires entered and gradually destroyed extensive areas of surrounding forest (Wade and McVean 1969; Paijmans and Löffler 1972). In most years scrub bordering the forest acts as a fire break protecting it from fire damage (Wade and McVean 1969; Gillison 1970; Plate 40), as also may a local belt of lichens around the forest (Coode and Stevens 1972).

It is also likely that many sedge-grass swamps were formerly forested. The local presence of lower montane swamp forest indicates that permanent swampiness need not prevent forest growth. The author has found charred logs on the swampy slopes of the Neon Basin at 2800 m west of Mt Albert Edward, and soil augering in sedge swamps is often hampered by buried tree trunks (Haantjens, pers. comm.).

Space that became available through forest destruction was filled by grasses and herbs. Most of these have probably drifted down from adjacent natural grasslands above the tree line, because an upward spread of lowland grasses was, and still is, hampered by a belt of forest in most localities. This may be one of the reasons why the grasslands above 2500 m are floristically more related to upper montane than to lower grasslands.

Except possibly on Mt Wilhelm, where deliberate burning was officially banned twenty years ago, parts of all high-mountain grasslands are fired each year. The long-term effect, as in the lowlands, is that only the most fire-resistant shrubs, perennial grasses and herbs can survive. However, fires are much more irregular than in lowland grassland, because of the general greater dampness, shorter and less reliable dry season, and more numerous permanently wet and hence less flammable patches. The short-term effect of fires produces a mosaic of successional and subclimax stages. Some pioneer annual herbs and grasses complete their life cycle shortly after a burn. Others attain their maximum growth and flowering peaks later, and may in turn be suppressed, possibly after some years, by a more or less stable subclimax of *Gleichenia*. Certain fire-resistant sedges, notably *Machaerina rubiginosa* and *Rhynchospora rugosa*, are also important components of the succession (Gillison 1969). Thus in defining grassland units very finely, as Wade and McVean (1969) have done for Mt Wilhelm, one runs the risk of describing 'associations' that are in fact only seral stages.

The partly haphazard species distribution, together with the wide environmental tolerance of many grassland components may be the reason for the different and, at least for some of the dominants, conflicting findings of various authors. For instance, the *Deschampsia klossii-Danthonia penicillata* association of Wade and McVean (1969: 82), which is dominated by *D. klossii*, is absent in areas of impeded drainage on Mt Wilhelm, whereas Walker (1968) on the same mountain found high frequency values for *D. klossii* on swampy plots, and concluded that tolerance of waterlogging assures the success of *D. klossii* on swampy sites. On Mt Wilhelm and Mt Giluwe *D. klossii* is seldom if ever observed on ground recently disturbed by burning (Wade and McVean 1969: 88), but Brass (1941B: 322) reported that on Mt Trikora grassland dominated by *D. klossii* developed 'as a distict ecological unit on well drained soil . . . on areas patently deforested by fire'. Hoogland (1958: 221) noted that very few grassland species on Mt Wilhelm were restricted to only one of two quite distinct aspects, though their frequency might be quite different.

Tree fern savanna probably indicates formerly forested terrain, a view accepted by Brass (1941b: 331) and Wade and McVean (1969: 59), and supported by air photo patterns (Plate 42). Wardle (1971: 398), however, believes it may well be the natural vegetation in many places. In some respects the New Guinea tree fern savanna resembles the paramos of the South American Andes and East African mountains. However, the paramos are at higher altitudes and under more severe climatic conditions, and are an undoubted natural vegetation type.

The beginnings of a succession from grassland back to forest can occasionally be observed, starting at scrub bordering forest and at the edges of shrubberies surrounded by grassland. Similar successions have been observed on Java by van Steenis (1968), who compared the process with the closing of a wound on man's skin. Conifers are usually prominent in the regeneration zones, as they are light-demanding, frost-hardy, and drought-resistant. If a fire ban, as imposed on Mt Wilhelm, proves to be effective, it will be possible, in due course, to determine to what extent frost is responsible for retarding the process of forest regeneration.

Man-made vegetation

As in the lowlands, abandoned gardens are first invaded and soon covered by various weeds and grasses, including *Ageratum, Stachytarpheta, Arthraxon ciliaris* and *Ischaemum polystachyum*. These are followed, usually within a year, by herbaceous and woody creepers and climbers, tree ferns, shrubs, and the grasses *Imperata cylindrica* and *Miscanthus floridulus*. *Imperata* and *Miscanthus* are the strongest competitors. Absence of burning and presence of rooting pigs favour *Miscanthus* (Walker 1966) and, provided there is no regardening, young secondary forest develops via a *Miscanthus*-dominated stage. On the other hand, repeated burning results in a more or less stable grassland initially dominated by *Imperata* (Robbins 1963).

Trees such as *Castanopsis, Lithocarpus, Sloanea, Elaeocarpus, Euodia, Engelhardtia* and in places *Nothofagus* are prominent in advanced regrowth. Usually these trees are present in a mixture, although the oaks and *Engelhardtia* also form pure stands.

Resources

The lower montane forests are generally richer in timber volume than the lowland hill and mountain forests (Lane-Poole 1925), owing mainly to the abundance of *Nothofagus* and the presence of various conifers. However, the generally rugged terrain restricts their utilisation. A number of sawmills, notably in the Highlands, provide output for a growing local market, where the demand generally exceeds supply (Papua New Guinea Department of Forests 1973). Favoured timber trees include *Nothofagus, Calophyllum, Syzygium, Dryadodaphne* and conifers.

In the Highlands plantations of *Casuarina papuana, Eucalyptus deglupta,* and other fast-growing species help to overcome the great shortage of firewood and small timber for home use. They will also serve in the burgeoning tea industry, which has recently expanded its tea-growing area to include drained swamp land.

Timber exploitation of lower montane forest is generally not followed by regeneration or re-afforestation measures, and in the highest parts of the zone much forest has been and is being destroyed by fire. There is therefore an urgent need for protection of the high-mountain environment through forest reservation and the imposition of fire bans.

The Upper Montane Zone

Vegetation and habitat

The upper montane zone consists of discontinuous areas of ridges, peaks and plateaus above 3000 m on mainland Papua New Guinea, several of them rising to over 4000 m (Plate 43). The climate is generally drier than in the zone below, although the sun may not break through the clouds, mist and rain for periods lasting several weeks. During the dry season many days are clear and the nights are cold, with frost towards dawn; dry spells of two or three days are common, and some last for two weeks or longer.

The main vegetation types, forest and grassland, grow in mosaic. With increasing altitude the proportion of grassland increases as the area of forest is reduced to islands in a sea of grass. Herbaceous swamp communities occupy shallow depressions within grassland throughout the upper montane zone, and with increasing altitude

Plate 43
Garden with taro (*Colocasia esculenta*) and bananas (*Musa* spp.) on 44° slope, Bougainville Island.

become progressively more common also on slopes.

The borders between grassland and forest in the upper montane and highest parts of the lower montane zones are marked by a discrete community consisting mainly of light-demanding shrubs which find their optimum habitat here (Wade and McVean 1969; Kalkman and Vink 1970; Gillison 1970; Plate 40). Most members of the transition community are adapted to both drought and fire, and many seed profusely. Many also propagate by falling over and suckering, a characteristic that enables them to move backwards or forwards with the forest edge. Some species change form under an advance or retreat of the transition with varying fire frequency. *Mussaenda scratchleyi*, for example, occurs as a compact shrub with large leaves in open grassland, but as a small-leaved canopy climber in the forest side of the transition. *Nepenthes mirabilis* occurs as an erect herb in the transition front, but towards the forest becomes a canopy climber up to 10 m high with fewer and differently formed pitchers (Gillison 1971).

The limit of tree growth is generally at about 3900 m but varies much from place to place, like other altitudinal vegetation boundaries. The highest sites with trees are in the most sheltered positions on the warmest and driest soils (Brass 1941b).

As the tree limit is approached most trees become low-branched and shrub-like, and the distinction between forest and scrub becomes vague. At the tree limit the vegetation assumes a parkland aspect, and consists of either dense shrubberies alternating with grassland, or low forest opening out into woodland with an undergrowth of shrubs, ferns, grasses and sedges.

Above the tree line shrubberies and single shrubs gradually

decrease in height and frequency, and above 4400 m rapidly peter out. Rosette and cushion herbs, mosses, lichens and low ferns become progressively more abundant with increasing altitude and largely replace the grasses above 4300 m. The proportion of bare ground, unsuited to plant growth because of either lack of soil or extreme steepness, becomes progressively higher.

Plant life, including some grasses, occurs to the highest altitudes where soil is present. Grasses have been recorded under rocks at 4500 m on Mt Trikora (Brass 1941b), and next to the edges of glaciers at 4600 m on the Jaya Mountains by Hope (1973), who believes that a true climatic limit to plant growth is not reached in New Guinea.

Upper montane pioneer communities cover many small areas on exposed rock, landslips, and moraines, where they grow in direct contact with parent rock material. One reason for the great variety in the floristic make-up of these communities is probably that the high mountains of Papua New Guinea are markedly different in rock type. For instance, Mt Albert Edward and Mt Victoria are built up of mica schist, on Mt Wilhelm the rock type is granidiorite, the summit areas of the Saruwaged Range (Mt Bangeta) are mainly limestone, and Mt Giluwe is an extinct volcano of mafic lava.

Most high-altitude shrubs have a prostrate habit and small, stiff, leathery, appressed and often shiny leaves with inrolled edges. Their xerophytic appearance may be an adaptation to the combined effects of drying winds and periodic low atmospheric humidity, while shiny leaves are probably a defence mechanism against strong light intensity. The upper montane forests grow at too high an altitude to be of commercial value, but fulfill an useful function in the protection of watersheds. However, not only the forests, but the upper montane environment as a whole needs to be preserved for its unique flora and fauna.

Vegetation types

Forest. The canopy of upper montane forest ranges from about 18 m in the lower part of the zone to about 6 m at the forest limit, but at similar altitudes within this range varies greatly owing to regional differences in floristics and local differences in topography. The forest is made up of densely growing trees with commonly thin, crooked trunks, although the conifers usually have thicker and straight stems.

In the lower part of the zone the forest is relatively rich in species. Trees in and below the canopy belong to the families Myrsinaceae, Ericaceae, Myrtaceae, and Rubiaceae. The conifers *Podocarpus, Dacrycarpus, Papuacedrus,* and large-leaved *Schefflera* emerge some 5 m above the canopy. Tree ferns feature in the shrub layer, but pandans and climbing bamboo are absent. Among the most common shrubs are the Rubiaceae *Amaracarpus* and *Coprosma.*

On steep stony slopes where there is little soil, upper montane forest is very low and shrubby, and resembles scrub above the tree line, except that it is richer in species and not quite so dwarfed. Common on such sites are *Papuacedrus, Coprosma, Xanthomyrtus,*

Ericaceae, and Epacridaceae.

Near the tree limit most trees are low and shrub-like, although some conifers, notably *Dacrycarpus compactus,* retain their true tree stature. Other structural changes are the gradual disappearance of lianes, and the decreasing abundance of epiphytes. Relatively few woody genera make up the forest, mainly *Coprosma, Drimys, Olearia, Pittosporum, Rapanea, Rhododendron,* and *Vaccinium.*

Grassland. The structure and composition of upper montane grassland are dependent mainly on altitude, soil depth, drainage conditions and, particularly in the higher regions of the zone, exposure. Mid-height tussock grasses predominate on relatively well drained sites with deep soils, and continue to well above the tree line on favourable spots. Low grasses predominate on sites that have shallow soils or are poorly drained; such sites become more common with increasing altitude. Low grasses may consist mainly of stunted tussock grasses such as *Deschampsia klossii,* or may form a more or less continuous sward of dwarf grasses such as *Monostachya oreoboloides* and species of *Poa, Festuca* and *Danthonia. M. oreoboloides* and tiny rosette herbs such as *Plantago* also densely cover walking tracks and, within mid-height grassland, act as fire breaks.

Rosette and cushion herbs, mosses, lichens and ferns abound, and in places have a higher coverage than the grasses. *Gleichenia vulcanica* is a regular dominant up to 4000 m on both very wet sites and on rocky slopes with little soil. The curious finger fern *Papuapteris linearis* with erect, stiff and narrow leaves commonly dominates with tussock grasses on Mt Wilhelm above 4000 m (Wade and McVean 1969). Shrubs are present almost throughout upper montane grassland, but become generally scarce and mostly dwarfed to less than 20 cm above 4000 m, although genera such as *Coprosma* and *Drimys* on favourable sites at this altitude still reach a height of over 2 m and patches of low heath occur above 4000 m on Mt Wilhelm.

Rather surprisingly, upper montane grassland does not show a decrease in number of species with increasing altitude, in contrast to forest. Wade and McVean (1969: 82) report on Mt Wilhelm that 'species numbers per unit area are highest around 3400 m, beyond which there is a slight fall to 3900 m and another slight increase above this'. Brass (1941b: 328) recorded a steady gain in the total number of herbs and particularly of Compositae and grasses up to about 4000 m on Mt Trikora; above this elevation species diminished rapidly in number and abundance.

Herbaceous swamp vegetation. Herbaceous communities consisting of a mixture of low herbs, sedges, grasses, and mosses occupy depressions, fringe open water and, in the higher parts of the zone, also occur on slopes. The plants grow in waterlogged peat, which is either solid or more or less liquid and floating. Herbs, grasses, and mosses are prominent in relatively great variety in shallow bogs, and scattered dwarf shrubs are also present. In deeper bogs sedges predominate, and shrubs are virtually absent. Horsetails (*Equisetum*) may occur in moving water.

Among the commonest grasses are *Anthoxanthum angustum,*

Agrostis reinwardtii, and *Monostachya oreoboloides*. The sedge *Carpha alpina* and the fern *Gleichenia vulcanica* locally form pure stands. Herbs seldom determine the aspect, except the liliaceous *Astelia papuana*, which in many places grows in large gregarious cushions through which other herbs and some shrubs may protrude. Some of the commonest shrubs are *Leucopogon, Drapetes, Vaccinium*, and *Trochocarpa*.

Vegetation History*

The vegetation types described above in terms of floristics, structure and successional status have developed over a considerable period of time. They comprise floristic elements of both southern and northern affinity (considered in detail in Part I), which probably entered within Tertiary and early Quaternary times (Walker 1972), together with others developed within the country itself. Virtually no palynological data are available for these early periods; the present level of endemism suggests that there has been rapid speciation during the Quaternary, perhaps in response to changing environmental conditions.

The last 30,000 years or so of vegetation history for some parts of Papua New Guinea have recently been partially documented (Flenley 1967, 1972; Walker 1970, 1973; Powell 1970a, 1970b; Hope 1973). Pollen analytical studies have concentrated on highland areas where problems inherent in attempting analyses of present and past tropical vegetation (Walker 1970) are probably less difficult to overcome than elsewhere.

Altogether five sites in the Western Highlands (Draepi, Manton) and Enga Districts (Inim, Birip, Sirunki), ranging in altitude from 1590 m to 2550 m have been studied, together with a series of sites on Mt Wilhelm lying between 2740 m and 4420 m altitude. The time period covered by the Enga and Western Highlands sites is from beyond 32,000 years ago to the present, with a break between 18,000 and 5000 years ago in the Draepi site. The Mt Wilhelm sites record 22,000 years of history. A few samples from older deposits in the Tari-Koroba area of the Southern Highlands District have also been analysed (Williams *et al.* 1972).

The pollen diagrams from all sites provide evidence of vegetation changes which can be interpreted in climatic and anthropogenic terms. The following summary is based largely on unpublished data and tentative interpretations, which will undoubtedly be altered when further studies are undertaken; details of pollen diagrams are to be found in the studies cited above.

The Tari-Koroba samples register the presence of upper montane grassland, coniferous forest and mixed forests (Epacridaceae, Elaeocarpaceae, Myrtaceae, *Quintinia, Ilex*) between (and older than) 32,700 years B.P. and 38,600 years B.P. at sites today at 1500 m altitude (Williams *et al.* 1972). 150 km to the east-north-east, near Mt Hagen, the Draepi site records upper montane grassland and mixed forest as being more important than beech forest

---

* This section was written by Dr J. M. Powell, together with Dr G. S. Hope, Research School of Pacific Studies, Australian National University, Canberra.

(*Nothofagus*) from possibly 38,000 years B.P. to about 30,000 years B.P. (Powell 1970a). Beech forest then becomes dominant and is particularly well represented between 23,000 and 18,000 years ago. At Sirunki (2500 m) upper montane vegetation (indicated by *Astelia, Drapetes, Potentilla, Ranunculus, Styphelia, Coprosma*) was present between 16,000 years B.P. and 12,000 years B.P. (Walker 1970) and on Mt Wilhelm upper montane grassland and tree ferns (*Cyathea*) are present at altitudes below 2740 m until about 10,200 years B.P. (Hope 1973). At Lake Inim (2550 m) upper montane vegetation is present locally until less than 12,000 years ago (Flenley 1967, 1972).

The presence of upper montane species at 1500 m over 30,000 years ago and at 2550 m to perhaps 10,500 years ago implies a considerable shift in vegetation zones, of between 500 m and 1500 m, and this can only be explained in terms of climatic change. For Mt Wilhelm Hope (1973) suggests that the minimum depression of local communities near the site at 2740 m was about 700 m, while the tree line was lowered by 1200 m or more. With indubitable evidence for glaciation of New Guinea's high mountains dating to before 10,000 B.P. (Löffler 1972; Hope and Peterson 1974; Galloway *et al.* 1973) the vegetation changes recorded may be related to the temperature changes which accompanied the waxing and waning of the ice.

From 12,000 years B.P. to 5000 years B.P. at Sirunki beech and mixed forest grew on the slopes surrounding the site, and at times oak forest (*Castanopsis*) was also important. At Inim beech dominated the forest from 8000 years B.P. until about 2000 years ago (Walker 1970). On Mt Wilhelm mixed montane forests with some oak grew near 2740 m after 10,200 years B.P. and from about 8600 years B.P. oak forest assumed greater importance. Upper montane forest (*Rapanea, Dacrycarpus*) reached 4000 m altitude at 8300 years B.P. Hope (1973) has interpreted the rising forest tree line in terms of ameliorating climatic conditions; by 8300 years B.P. the climate was milder than at present.

At Inim and Sirunki forests were probably distributed much as they are today; the increase of oak at different periods at Sirunki could be interpreted in terms of slightly milder or perhaps drier conditions. Changes in *Castanopsis* values about 6500 years B.P. in Mt Wilhelm pollen diagrams have been interpreted in this way by Hope (1973) but he suggests also that anthropogenic influences could have been important at that time.

Certainly by 5000 years B.P. at Draepi there was considerable forest reduction associated with an increase in light-demanding woody taxa and grassland species, and this has been interpreted in terms of forest clearance by man (Powell 1970a, b). The oak forest bordering the Manton site (1590 m altitude) was very disturbed by then also and at both sites more direct evidence of human activities (ditches, digging sticks) is present and dated to about 2300 years B.P. This is considered in more detail later in this volume by Powell. Reduction of forest interpreted as due to anthropogenic influences began about 2000 years B.P. at Birip (1900 m) and 1600 years ago at Lake Inim according to Flenley (1970). The forest was somewhat disturbed at both sites before those dates, however, and at Sirunki disturbance dates to 4500 years B.P. (Walker 1970). There is no direct evidence at present to associate these disturbances with man although such an interpretation would appear to be reasonable.

By 5000 B.P. on Mt Wilhelm grassland had replaced forest at 3910 m altitude but forests at lower altitudes (3550 m, 2740 m) were unaffected; Hope (1973) has interpreted this tree line depression in terms of climatic deterioration, which also caused the appearance of ice at still higher altitudes. More recently destruction of both lower and upper montane forest is indicated and this is considered to be due to human activities. The decline in lower montane forest from 1100 years B.P. is associated with a steady increase in *Casuarina* and *Trema* and may be considered indicative of garden expansion in the Chimbu valley. At 3510 m altitude the replacement of forest by grassland and tree ferns from 800 years B.P. may be attributed to burning and cutting by man; the tree line is stable at this time, at about 3900 m altitude.

The changes described and interpreted above all relate to vegetation growing at or above 1500 m altitude. At present no data are available for lowland and mid-mountain areas. While it would be foolish to extrapolate the present limited data too far, the degree of vegetational shift indicated for the highlands during the last glacial period implies there was at least some modification of lowland and mid-mountain vegetation distribution patterns. Kershaw's (1970, 1971) vegetation history studies in North Queensland indicate replacement of former sclerophyllous forest by present day tropical rain forest about 7000 years ago; his work perhaps foreshadows the results of future Papuan lowland studies.

The limited evidence available so far is consistent with the following:

38,000-30,000 years B.P.	Extensive upper montane communities, tree line at 1900-2100 m. Lowland rain forest in Queensland gives way to sclerophyll types. Arafura plain is exposed and presumably arid (Nix and Kalma 1972).
30,00-12,000 years B.P.	Tree line at about 2000-2300 m. Beech forest prominent across the island, semi-arid vegetation possibly continuous along southern New Guinea and in the Arafura area.
12,000-8500 years B.P.	Rapid rise in tree line to 4000 m and replacement of some highland beech forests by mixed lower montane and oak forests.
8500-5000 years B.P.	Possibly higher tree line and milder climate than present. Vegetation disturbance by man possibly begins. Sea level rises and the coastline assumes present position; lowland rain forest spreads over southern New Guinea. Broad delta and swamp forests develop.

5000 to present	Some climatic deterioration caused tree line lowering and perhaps affected forest composition. Widespread forest destruction and clearance by man, possible extension of lowland grassland and savanna by burning. Extension of riverine swamps as rivers aggrade.

It can be suggested that the present distribution of the vegetation of New Guinea has developed over the last 10,000 years, except in core areas of middle to low altitude around the flanks of the central mountain chain. Even within that time minor fluctuations of climate have caused changes in the distribution and these continue at present. Moreover, considering that the present vegetation was probably well adapted to the somewhat different Pleistocene cool conditions (which lasted for longer than 30,000 years), then some puzzling distribution patterns and species behaviour could possibly be explained in terms of insufficient time being available for adaptation to present environmental conditions.

The Vegetation of Irian Jaya

The vegetation of the western half of the island of New Guinea appears, from various reports, to be similar to that of the eastern half. In both parts about three-quarters of the total area is under forest, but in the west the absence of volcanic activity results in a somewhat higher proportion of climax forest. Fires have been less devastating, at least in the upper regions of Mt Trikora, than on most high mountains of Papua New Guinea, probably because of greater wetness (Brass 1941b).

As in Papua New Guinea, mangroves are more extensive along the south coast than along the northern shores. Monsoonal savannas and dry evergreen forest roughly cover a triangular area stretching along the coast from the mouth of the Fly River in Papua New Guinea to beyond Merauke in Irian Jaya, and having its top where the Fly forms the international border. The eucalypt savanna of the central-south coast region of Papua New Guinea is not present in Irian Jaya.

The accounts of the ascents of Mt Doorman, 3580 m, by Lam (1945), and of Mt Trikora, explored to 4500 m by Brass (1941b), show that the same altitudinal changes in the structure and floristics of the forest are present as in Papua New Guinea. Forest surveys (unpublished reports) indicate that conifers in general, and especially *Agathis*, are more common, but that *Araucaria* is much scarcer than in Papua New Guinea. *A. hunsteinii* has not been recorded in Irian Jaya (Gray 1973). As in the east, *Nothofagus* is a common dominant in lower montane forest. On Mt Trikora conifers drop out above about 3500 m and are replaced by *Aralia* spp. (Brass 1941b), but on Mt Doorman and Mt Jaya, as also generally in Papua New Guinea, one or more conifers reach to the limit of tree growth. A species of *Casuarina* dominates the forest on Mt Doorman from about 2750 m upwards, and according to Lam reaches its upper limit not far from 4000 m altitude. Such high-altitude forest and woodland of *Casuarina* has not been recorded elsewhere in New Guinea (Hope, in press). The tree fern savanna described for Papua New Guinea is one of the most extensive vegetation types in the Jaya Mountains and is also present on Mt Trikora.

The upper limit of forest on Mt Jaya, at 3900 m, is the same as on the mountains in Papua New Guinea. As mentioned earlier, its interpretation is rather subjective and depends on where one begins to call an increasingly low and crippled forest a scrub. Most species that are common throughout the high-altitude grasslands of Papua New Guinea, such as *Deschampsia klossii, Astelia papuana,* and *Styphelia suaveolens,* are also widespread in Irian Jaya, although their proportion and altitudinal range may be different. Where floristics have been examined and compared in detail, similarities became less obvious. Of the twenty-nine vegetation units, including twenty-one 'associations', described from Mt Wilhelm above 3100 m (Wade and McVean 1969) only eight had close analogues on Mt Jaya (Hope, in press).

Differences between the high-mountain vegetations of Papua New Guinea and Irian Jaya are probably partly due to the climate, which is progressively wetter from east to west. Thus, in Papua New Guinea, Mt Albert Edward is drier than Mt Wilhelm, and both are drier than the Jaya Mountains in Irian Jaya, which have an exceptionally wet and mild climate (Hope, in press). The distribution of grassland dominated by *Danthonia archboldii,* for instance, may well be related to this climatic trend. This type of grassland is common on Mt Albert Edward (Paijmans and Löffler 1972: 61), is very restricted on Mt Wilhelm (Wade and McVean 1969: 88), rare on Mt Trikora (Brass 1941b: 329), and absent on Mt Jaya (Hope, in press).

Tall closed shrubland above the tree line is a vegetation type very scarce in Papua New Guinea. It is, however, common on Mt Jaya and forms a complete cover over its altitudinal range on Mt Mandala. It is possible that this type has been destroyed by fire on many New Guinea mountains, and persists only on the most inaccessible summits, although climate may also play a part (Hope, in press).

Timber volumes and number of stems per unit area compare favourably with those of similar forest types in Papua New Guinea (Richardson 1970). Over large areas merchantable volumes approach 100 m^3 per ha, a very high figure by Papua New Guinea standards. For both areas average figures are probably not greatly different from those for mixed forest in Indonesia, Malaya and the Philippines which on average contain between 60 and 120 m^3 of merchantable timber per ha, but are richer in dipterocarps. Irian Jaya lags behind in commercial utilisation of its forest resources, and notions of conservation, sustained-yield management and re-afforestation, which are beginning to be implemented in Papua New Guinea, are yet to be developed in the west.

PART III

Ethnobotany

J.M. Powell

Introduction

Throughout New Guinea plant resources form the basis of life whether the people are nomadic hunters and gatherers, fishermen, horticulturalists or sedentary agriculturalists. Plants provide them with food, medicines and poisons, with the raw materials needed to build houses and shelters, canoes and rafts and to make tools and weapons, clothing and containers. When man first reached New Guinea many thousands of years ago he found an island extremely diverse in environments and vegetation, ranging from the lowland coastal and riverine swamps with mangrove, nipa and sago palm forests, through dry plains covered with grassland and savanna to luxuriant lowland, foothill and high mountain rain forests. Plant resources probably far outweighed animal resources, the New Guinea flora comprising perhaps 9000 species in 1465 genera and 246 families (Part I). Some of the food resources, for example wild yams and bananas, may have been known to the early immigrant, while others required experimentation and perhaps technological improvisation before they could be used. Domestication and selection of cultivars suited to particular environments occurred over a period of time and later immigrants introduced new crops domesticated elsewhere; these were accepted and utilised alongside the local resources. The type of subsistence pattern developed in any area depended heavily on the environmental conditions present and resource availability may have frequently limited population development.

In this part the ethnobotany of the New Guinea people is discussed with regard to the staple and supplementary crops, the agricultural systems present and the wild plant resources available for utilisation as food and as raw materials for products of material culture. The data available in the literature are extremely uneven and inadequate; some areas have been studied in far more detail than others; many of the studies consulted have been undertaken by anthropologists and geographers and the ethnobotanical aspects have not been a major part of their work; early workers usually cite indigenous plant names and botanical identifications are lacking. Some data have been brought together, however, by Massal and

---

I should like to thank Dr A. Chowning, Dr R. Hide and Mr J.S. Womersley for allowing me to use unpublished data and Professor D. Walker, Dr G.S. Hope, Dr J. Allen and Dr D. Yen for their useful comments and discussion on the text. I am grateful for the assistance of Messrs A. Kulunga, Moge and Andiki Dibaja, among many others, with field recording in the Mt Hagen and Tari areas of the Highlands.

Figure 3.1 Locations of specific studies consulted:

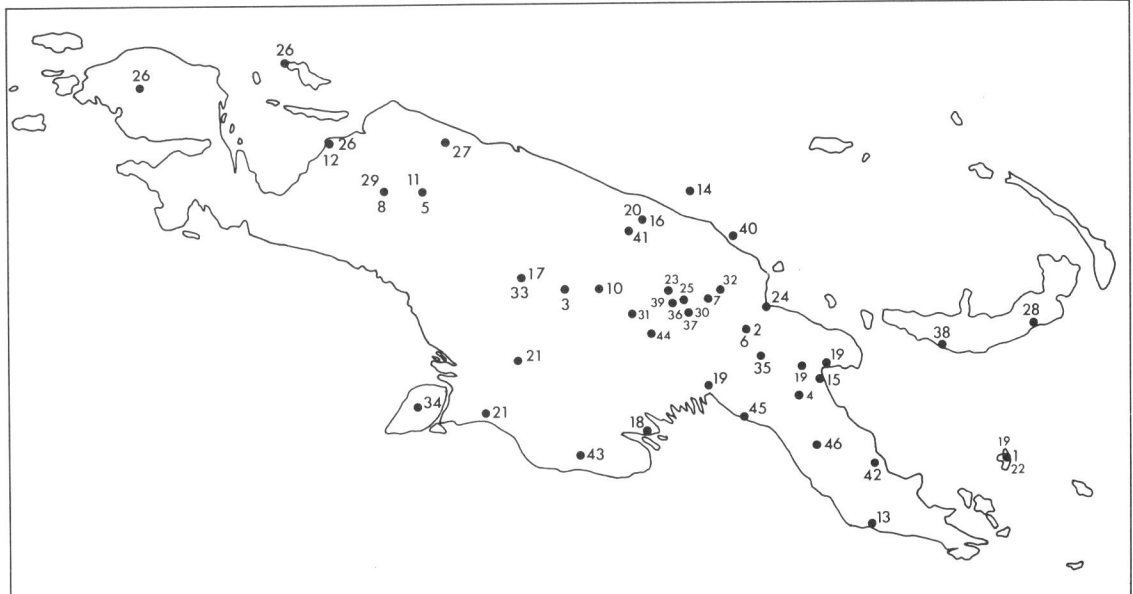

1 Austen 1945	17 Kooijman 1962	31 Powell 1973
2 Bailey and Whiteman 1963	18 Landtman 1933	32 Rappaport 1967
3 Barth 1971	19 Langley 1950, in Hipsley and Clements 1950	33 Reynders 1962
4 Blackwood 1939, 1940, 1950		34 Serpenti 1965
5 Brass 1941a	20 Lea 1965, 1966	35 Sorenson and Gajdusek 1969
6 Brookfield and Brown 1963	21 Luyken and Luyken-Koning 1955	36 Stopp 1963
7 Clarke 1971		37 Strathern, A. and A.M. 1971
8 Couvée *et al*. 1962	22 Malinowski 1918, 1935	38 Todd 1934
9 Floyd 1954	23 Meggitt 1957, 1958a, b	39 Waddell 1972
10 Hatanaka and Bragge 1973	24 Miklouho-Maclay 1885	40 Wedgwood 1934
11 Heider 1970	25 Oomen and Corden 1970	41 Whiting and Reed 1939
12 Held 1957	26 Oomen and Malcolm 1958	42 Williams 1930
13 Hipsley and Kirk 1965	27 Oosterwal 1961	43 Williams 1936
14 Hogbin 1938	28 Panoff 1970a, b, 1972	44 Williams 1940a
15 Hogbin 1951	29 Pospisil 1963	45 Williams 1940b
16 Kaberry 1941	30 Powell 1970a, b, 1974a, *et al.* 1975	46 Williamson 1912

Barrau (1956), Barrau (1958), Treide (1967), Straatmans (1967, 1971), Hide (1974) and Lea (1975). The locations of specific studies consulted are shown in Figure 3.1. The literature survey is supplemented by the author's own studies in the highland areas of Papua New Guinea.

Staple and Supplementary Crops

The staple foods are predominantly starchy, comprising tubers such as yams (*Dioscorea* spp.), and the sweet potato (*Ipomoea batatas*), corms such as taro (*Colocasia esculenta*) and related Araceae (*Cyrtosperma chamissonis* and the recently introduced *Xanthosoma sagittifolium*) and sago (*Metroxylon* sp.) and bananas

(*Musa* spp.). Supplementary crops, on the other hand, comprise a very wide range of green vegetables, fruits and nuts, such as *Hibiscus manihot,* the edible grasses (*Setaria palmifolia, Saccharum edule*) and spinaches (*Amaranthus hybridus, A. tricolor, A. viridis*), together with *Oenanthe javanica, Solanum nigrum, Rungia klossii,* sugarcane (*Saccharum officinarum*), breadfruit (*Artocarpus altilis*), coconuts (*Cocos nucifera*), *Barringtonia* spp., *Pandanus* spp., *Canarium* spp. and *Terminalia* spp. among others. The main species recorded as food in New Guinea today are listed in alphabetical order of genera in Table 3.1.

Table 3.1 Plants used as food

c	= cultivated as supplementary food
cs	= cultivated as staple food
cw	= wild form used as staple food
t	= transplanted
w	= wild form used as supplementary food
*	= introduced
EHD	= Eastern Highlands District

Plant name	Family	Part eaten	Area
Abroma angustum (c)	Sterculiaceae	Leaves	Chimbu
Acalypha sp. (w,t)	Euphorbiaceae	Leaves	Jimi, Chimbu
Aglaia sp. (w)	Meliaceae	Fruit	Coastal
Albizia sp. (w)	Leguminosae	Leaves	Coastal
Aleurites moluccana (w,t)	Euphorbiaceae	Nut	Jimi, Chimbu
Allium spp. (c)	Liliaceae	Leaves	Widespread
Alocasia mocrorrhiza (c,w)	Araceae	Tuber, leaves	Widespread
Alpinia spp. (c,w)	Zingiberaceae	Leaves, rhizome	Widespread
Amaranthus hybridus (c)	Amaranthaceae	Leaves, young plant	Tari, Mt Hagen
Amaranthus tricolor (c)	Amaranthaceae	Leaves, young plant	Widespread, coastal
Amaranthus viridis (c)	Amaranthaceae	Leaves, young plant	Kainantu, highlands
Amomum polycarpum (c,t)	Zingiberaceae	Fruit	Jimi
Amomum sp. (w)	Zingiberaceae	Seeds	New Britain
Annona muricata (c)	Annonaceae	Fruit	New Britain, Trans-Fly
Arenga sp. (w)	Palmae	Heart 'cabbage'	Waropen
Arrhenechthites spp. (w)	Asteraceae	Leaves	Chimbu
Artocarpus altilis (c,w)	Moraceae	Fruit, seeds	Widespread
Artocarpus spp. (w)	Moraceae	Seeds, pulp	Fore, EHD, Jimi
Ascarina philippinensis (w)	Chloranthaceae	Leaves	Chimbu
Asplenium affine (w)	Aspleniaceae	Leaves	Watut, EHD
Asplenium sp. (w)	Aspleniaceae	Leaves	Watut, EHD
Astilbe sp. (w)	Saxifragaceae	Leaves, juice	Chimbu
Astronia sp. (w)	Melastomataceae	Young leaves	Watut, EHD
Athyrium esculentum (w)	Athyriaceae	Young fronds	New Britain
Bambusa spp. (c,w)	Poaceae	Shoots	Widespread
Barringtonia spp. (c,w)	Barringtoniaceae	Nut	Coastal
Begonia spp. (w)	Begoniaceae	Leaves, stems	Highlands
Beilschmiedia sp. (w)	Lauraceae	Leaves	Jimi
Bidens sp. (w)	Asteraceae	Seeds	Highlands
Blechnum sp. (w)	Blechnaceae	Young fronds	Wissel Lakes
Blumea riparia (w)	Asteraceae	Leaves	Chimbu
Brassica juncea (c)	Brassicaceae	Leaves	Jimi
Bruguiera conjugata (w)	Rhizophoraceae	Fruits	Waropen
Bruguiera eriopetala (w)	Rhizophoraceae	Fruits	Widespread
Burckela sp. (w)	Sapotaceae	Fruit	Coastal
Canaritum indicum (syn. *C. mehenbethane* and *C. commune*) (w,t)	Burseraceae	Nut	Coastal, New Britain
Canarium salomonense (w,t)	Burseraceae	Nut	Coastal
Capsicum frutescens (c)	Solanaceae	Fruit	Waropen, New Britain
**Carica papaya* (c)	Caricaceae	Leaves, fruit, flowers	Widespread
Caryota rumphiana (w)	Palmae	Pith of young trunk	New Britain

Plant name	Family	Part eaten	Area
Cassia sp. (w)	Leguminosae	Leaves	Coastal
Castanopsis acuminatissima (w,t)	Fagaceae	Nut	Jimi, Tari, Chimbu
Celosia argentea (c)	Amaranthaceae	Leaves	Chimbu
Chisocheton sp. (w)	Meliaceae	Nut	Jimi
Chloranthus officinalis (w)	Chloranthaceae	Leaves	Watut, EHD
Chloranthus sp. (w)	Chloranthaceae	Leaves	Jimi
Cinnamomum sp. (w)	Lauraceae	Bark	Fore
Citrus spp. (w)	Rutaceae	Fruit	New Britain, Wissel Lakes
Cocos nucifera (cs,c,w,)	Palmae	Nut, milk	Widespread lowlands, coastal
Coix gigantea (w)	Poaceae	Stems	Highlands
Coix lachryma-jobi (w)	Poaceae	Seeds	Tari
Coleus scutellarioides (c,w)	Lamiaceae	Leaves	Jimi, New Britain
Coleus sp. (w)	Lamiaceae	Leaves	Watut, EHD
Colocasia esculenta (cs,c)	Araceae	Corm, leaves	Widespread
Commelina cyanea (c)	Commelinaceae	Leaves, shoots	Jimi
Commelina diffusa (c,w)	Commelinaceae	Leaves, shoots	Tari, Chimbu
Convolvulus sp. (w)	Convolvulaceae	Leaves	Coastal
Cordyline terminalis (c)	Liliaceae	Tuber	
Crotalaria linifolia (w)	Leguminosae	Leaves	Chimbu
Ctenitis sp. (w)	Aspidiaceae	Young fronds	
*Cucumis sativus (c)	Cucurbitaceae	Fruit	Widespread
*Cucurbita maxima (c)	Cucurbitaceae	Fruit, leaves	Widespread
*Cucurbita moschata (c)	Cucurbitaceae	Fruit, leaves	Jimi
*Cucurbita pepo (c)	Cucurbitaceae	Fruit, leaves	Widespread
Curcuma domestica (c,w)	Zingiberaceae	Rhizome	Widespread
Cyathea angiensis (c,w)	Cyatheaceae	Fronds	Jimi
Cyathea contamiana (w)	Cyatheaceae	Fronds	New Britain
Cyathea rubiginosa (w)	Cyatheaceae	Fronds	Jimi
Cyathea spp. (c,w)	Cyatheaceae	Fronds	Jimi, Wissel Lakes, Tari, Mt Hagen, Kainantu
Cyclosorus truncatus (w)	Thelypteridaceae	Fronds	Jimi, New Britain
Cynoglossum sp. (w)	Boraginaceae	Leaves	Chimbu
Cypholophus sp. (w)	Urticaceae	Leaves	Watut, EHD
*Cyphomandra betacea (c)	Solanaceae	Fruit, leaves	Widespread highlands
Cyrtandra sp. (w)	Gesneriaceae	Leaf	Widespread
Cyrtosperma chamissonis (c,w)	Araceae	Tuber	Jimi, New Britain
Decaspermum sp. (c,w)	Myrtaceae	Leaves	Chimbu
Dennstaedtia spp. (w)	Dennstaedtiaceae	Fronds	Jimi, New Britain, Wissel Lakes
Desmodium microphyllum (w)	Leguminosae	Leaves	Watut, EHD, Chimbu
Desmodium repandum (w)	Leguminosae	Leaves, seeds	Chimbu
Dioscorea alata (cs,c)	Dioscoreaceae	Tuber	Widespread
Dioscorea bulbifera (c, w)	Dioscoreaceae	Tuber, bulbils	Widespread
Dioscorea esculenta (cs, c)	Dioscoreaceae	Tuber	Widespread
Dioscorea hispida (c,w)	Dioscoreaceae	Leaves, tubers	New Britain
Dioscorea nummularia (c)	Dioscoreaceae	Tuber	Widespread
Dioscorea pentaphylla (c, w)	Dioscoreaceae	Tuber	Widespread
Dioscorea spp. (w)	Dioscoreaceae	Tuber	New Britain, Jimi
Diospyros spp. (w)	Ebenaceae	Fruit	Fore, coastal
Diplazium ?asperum (w)	Athyriaceae	Frond	Watut
Diplazium cordifolium (w)	Athyriaceae	Frond	Watut, EHD
Diplazium sp. (w)	Athyriaceae	Frond	Jimi
Dolichos lablab (c)	Leguminosae	Leaves, seeds	Jimi, Tari, Chimbu
Dryopteris cf. *arbuscula* (w)	Aspidiaceae	Frond	Watut

Plant name	Family	Part eaten	Area
Dryopteris sparsa (w)	Aspidiaceae	Frond	Watut, EHD
Dryopteris truncata (w)	Aspidiaceae	Frond	Watut, EHD
Elaeocarpus spp. (w)	Elaeocarpaceae	Nut	Highlands
Elatostema sp. (w)	Urticaceae	Leaves, stem tips	Highlands
Enhalus sp. (w)	Hydrocharitaceae	Leaves, stems	Waropen
Erythrina sp. (w,t)	Leguminosae	Leaves	Jimi
Euodia sp. (w)	Rutaceae	Leaves	Widespread
Ficus botryocarpa var *sub-albidoramea* (w)	Moraceae	Fruit	Jimi
Ficus copiosa (c,w)	Moraceae	Fruit, young leaves	Tari, Chimbu
Ficus dammaropsis (c,w)	Moraceae	Leaves, fruit	Widespread highlands
Ficus iodotricha (w)	Moraceae	Leaves	Jimi
Ficus itoana (w)	Moraceae	Fruit	Jimi
Ficus nodosa (w)	Moraceae	Leaves	Jimi
Ficus pachyrachis (w)	Moraceae	Leaves	Jimi
Ficus pungens (w)	Moraceae	Leaves	Tari, Jimi
Ficus wassa (c,w)	Moraceae	Leaves, fruit	Jimi
Ficus spp. (c,w)	Moraceae	Leaves, fruit	Widespread
Finschia chloroxantha (w)	Proteaceae	Seeds	Highlands
Floscopa scandens (c,w)	Commelinaceae	Leaves, shoots	Tari
Garcinia spp. (w)	Clusiaceae	Fruit, leaves, bark	Wissel Lakes, Watut, EHD
Gleichenia spp. (w)	Gleicheniaceae	Young fronds	Highlands
Gnetum gnemon (c)	Gnetaceae	Leaves	Jimi, New Britain
Goodyera rubicunda (w)	Orchidaceae	Leaves	Watut, EHD
Graptophyllum pictum (w)	Acanthaceae	Young leaves	Watut, EHD, Chimbu
Gronophyllum chaunostachys (w)	Palmae	Heart cabbage	Jimi
Habenaria sp. (w)	Orchidaceae	Tubers	Markham
Hemigraphis sp. (c)	Acanthaceae	Leaves, young shoots	Jimi, Tari, Hagen
Hibiscus manihot (c)	Malvaceae	Leaves, shoots	Widespread
Hornstedtia lycostoma (w)	Zingiberaceae	Seeds	New Britain
Impatiens spp. (w)	Balsaminaceae	Leaves	Watut, EHD
Inocarpus edulis (w)	Leguminosae	Nut	Sepik, New Britain
Ipomoea aquatica (w)	Convolvulaceae	Leaves, shoots	Waropen
Ipomoea batatas (cs,c)	Convolvulaceae	Tuber, leaves	Widespread
Ipomoea reptans (w)	Convolvulaceae	Leaves, shoots	Waropen
Iresine herbstii (c)	Amaranthaceae	Leaves	Chimbu
Kaempfera galanga (c,w)	Zingiberaceae	Rhizome, leaves	Coastal, lowlands
*Lactuca sativa (c)	Asteraceae	Leaves	Widespread
Lactuca sp. (w)	Asteraceae	Leaves	Highlands
Lagenaria siceraria (c)	Cucurbitaceae	Leaves, young fruits	Widespread
Laportea sp. (w)	Urticaceae	Leaves	Watut, EHD
Lepidagathis sp. (c,w)	Acanthaceae	Leaves, stem tips	Kaironk
Leucosyke sp. (w)	Urticaceae	Leaf	Jimi
Lucinaea sp. (w)	Rubiaceae	Leaves	Chimbu
Lycopodium spp. (w)	Lycopodiaceae	Young shoots	—
Lysimachia japonica (w)	Primulaceae	Leaves	Chimbu
Madhuca spp. (w)	Sapotaceae	Fruit	Rai coast
Mangifera indica (c)	Anacardiaceae	Fruit	Coastal, lowlands
Mangifera spp. (w)	Anacardiaceae	Fruit	Some highlands, Jimi
*Manihot esculenta (c)	Euphorbiaceae	Tuber	Widespread
Metroxylon spp. (cs, ws, c, w)	Palmae	Pith (sago)	Widespread
Microcos sp. (w)	Tiliaceae	Leaf	Jimi
Morinda citrifolia (w)	Rubiaceae	Fruit, leaves	Coastal
Musa spp. (cs,c,w)	Musaceae	Fruit, stalk (famine)	Widespread
Myristica sp. (w)	Myristicaceae	Fruit	Waropen
Nasturtium officinale (w)	Brassicaceae	Leaves, shoots	Widespread highlands

Plant name	Family	Part eaten	Area
Nephrolepis biserrata (w)	Oleandraceae	Roots	Frederik-Hendrik Is.
Nypa fruticans (w)	Palmae	Seeds, heart cabbage	Coastal swamps
Ocimum basilicum (c)	Lamiaceae	Leaves	—
Oenanthe javanica (c,w)	Apiaceae	Leaves, shoots	Widespread highlands
Ophiorrhiza spp. (w)	Rubiaceae	Leaves	Chimbu
Palmeria sp. (w)	Monimiaceae	Leaves	Jimi
Pandanus brosimos (c,w)	Pandanaceae	Nut	Widespread highlands
Pandanus conoideus (c,w)	Pandanaceae	Fruit	Wissel Lakes, Jimi
Pandanus foveolatus (w)	Pandanaceae	Nut	Jimi
Pandanus julianettii (c,w)	Pandanaceae	Nut	Jimi, Hagen, widespread
Pandanus spp. (w)	Pandanaceae	Fruit, nut	Widespread highlands
Pangium edule (w)	Flacourtiaceae	Fruit	Rai coast
Papuacedrus papuana (w)	Cupressaceae	Leaves	Watut, EHD
Parartocarpus venenosa (c,w)	Moraceae	Fruit	New Britain
Parsonsia sp. (w)	Apocynaceae	Fruit	New Britain
Phaseolus lunatus (c)	Leguminosae	Leaves, pods, seeds	Jimi, Tari
**Phaseolus vulgaris* (c)	Leguminosae	Leaves, young pods	Widespread
Phrynium sp. (w)	Marantaceae	Young leaves	Jimi
Pilea spp. (w)	Urticaceae	Leaves, flowers	Watut, EHD
Piper stenocarpum (w)	Piperaceae	Leaves	Chimbu
Piper sp. (w)	Piperaceae	Leaves	Chimbu
Pipturus argenteus (w)	Urticaceae	Bark, leaves	Watut, EHD
Pittosporum pullifolium (w)	Pittosporaceae	Seeds	Highlands
Pollia sp. (c,w)	Commelinaceae	Leaves, shoots	Jimi
Polygonum chinense (w)	Polygonaceae	Leaves	Chimbu
Polypodium commutatum (w)	Polypodiaceae	Young fronds	Watut, EHD
Polypodium irioides (w)	Polypodiaceae	Young fronds	Watut, EHD
Polypodium linguaeforme (w)	Polypodiaceae	Young fronds	Watut, EHD
Polypodium spp. (w)	Polypodiaceae	Fronds	Jimi
Polyscias sp. (w)	Araliaceae	Young leaves	Jimi, New Britain
Pouzolzia hirta (w)	Urticaceae	Leaves, stems	Jimi
Procris sp. (w)	Urticaceae	Leaves	Chimbu
**Psidium guajava* (c)	Myrtaceae	Fruit	Biak Is., widespread
Psophocarpus tetragonolobus (c)	Leguminosae	Leaves, flowers, pods, seeds	Widespread
Pteris moluccana (w)	Pteridaceae	Root, young fronds	Watut, EHD
Pueraria lobata (c,w)	Leguminosae	Tuber	Highlands
Ranunculus pseudolowii (w)	Ranunculaceae	Leaves	Chimbu
Rhododendron gracilentum (w)	Ericaceae	Leaves	Chimbu
Rhodomyrtus novoguinensis (w)	Myrtaceae	Leaves, fruit	Watut, EHD
Riedelia carallina (c)	Zingiberaceae	Leaves	Chimbu
Rorippa sp. (c)	Brassicaceae	Young plant	Highlands
Rubus moluccanus (w)	Rosaceae	Fruit	Jimi, Chimbu
Rubus rosifolius (w)	Rosaceae	Fruit	Hagen, Tari
Rubus cf. *fraxinifolius* (w)	Rosaceae	Fruit	Wissel Lakes
Rungia klossii (c,w)	Acanthaceae	Leaves, shoots	Widespread highlands
Saccharum edule (c)	Poaceae	Young inflorescence	Widespread
Saccharum officinarum (c)	Poaceae	Cane sugar	Widespread
Saurauia sp. (w)	Saurauiaceae	Leaves	Watut, EHD
Scleria sp. (w)	Cyperaceae	Fruit	Tari
Selaginella opaca (w)	Selaginellaceae	Leaf	Chimbu
Setaria palmifolia (c,w)	Poaceae	Shoot	Widespread highlands
Sloanea archboldiana (w)	Elaeocarpaceae	Nut	Highlands
**Solanum melongera* (c)	Solanaceae	Fruit	Coastal
Solanum nigrum (c,w)	Solanaceae	Leaves, shoots	Widespread
**Solanum tuberosum* (c)	Solanaceae	Tuber	Widespread
Sonchus sp. (w)	Asteraceae	Leaves	Chimbu

Plant name	Family	Part eaten	Area
Spondias dulcis (w)	Anacardiaceae	Fruit	Coastal
Stellaria sp. (w)	Caryophyllaceae	Leaves	Chimbu
Stenochlaena sp. (w)	Blechnaceae	Young fronds	
Sterculia sp. (w)	Sterculiaceae	Nuts	Highlands
Symbegonia sp. (w)	Begoniaceae	Leaves	Chimbu
Syzygium aqueum (w)	Myrtaceae	Fruit	Wissel Lakes
Syzygium malaccense (c,w)	Myrtaceae	Fruit	Coastal, common
Syzygium sp. aff. *pachyclada* (w)	Myrtaceae	Fruit	Wissel Lakes
Syzygium spp. (w)	Myrtaceae	Fruit	Jimi
Terminalia catappa (w)	Combretaceae	Seeds	New Britain
Terminalia copelandii (c,w)	Combretaceae	Seeds	Coastal
Terminalia impediens (w)	Combretaceae	Seeds	Coastal
Terminalia kaernbachii (w)	Combretaceae	Seeds	Coastal
Terminalia solomonensis (w)	Combretaceae	Seeds	Coastal
Thelypteris sp. (w)	Thelypteridaceae	Frond	Wissel Lakes
Trichosanthes sp. (w)	Cucurbitaceae	Fruit	Jimi, Hagen, Chimbu
Trigonotis procumbens (w)	Boraginaceae	Leaves	Chimbu
Uncaria sp. (w)	Rubiaceae	Leaves	Chimbu
Viola betonicifolia (w)	Violaceae	Leaves	Chimbu, Kainantu
Wahlenbergia marginata (w)	Campanulaceae	Plant	Chimbu
*Xanthosoma sagittifolium (c)	Araceae	Corms	Widespread
*Zea mays (c)	Poaceae	Corn	Widespread
Zingiber zerumbet (c)	Zingiberaceae	Rhizome, leaves	Widespread
Zingiber spp. (w)	Zingiberaceae	Rhizome, leaves	Widespread

The coastal and lowland swamp dwellers

Extensive fresh water swamps are present along the north and south coasts of New Guinea, developed on the alluvial plains bordering the major rivers, the Fly, Kikori, Purari, Ramu and Sepik in the east, the Digoel, Eilanden and Mamberamo in the west. Mangroves occupy coastal mudflats and river estuaries. Population densities are low in such areas, usually two to four persons per km^2.

Sago as the staple. Sago (*Metroxylon* spp.) is the staple crop in these areas; its use as such has been recorded from Biak Island, the Waropen and Tor areas of the north coast of Irian Jaya and from the Sepik in Papua New Guinea (Oomen and Malcolm 1958; Held 1957; Oosterwal 1961; Whiting and Reed 1938; Townsend 1971) and from the Mandobo and Marind areas of the south coast of Irian Jaya, the Fly River and Gulf Districts of Papua New Guinea (Reynders 1962; Oomen and Malcolm 1958; Riley 1923; Williams 1936, 1940a).

The sago palm grows 10-18 m tall in natural stands in fresh water swamps; it may be tended and seedlings and shoots are thinned and often transplanted. The plant takes 8-15 years to mature. It is usually felled just before flowering, as the starch stored in the trunk is at its maximum at this stage of the life cycle. To extract the sago the woody outer layer of the trunk is removed and the pith cut out with an adze-hafted pick, the cutting edge of which is made from obsidian, diorite, chert or other fine-grained stone or from hardwoods or a bamboo node. The pith is then carried to a waist-high trough made from the bark of the palm, or more often from a sago leaf petiole supported on other sago leaf petioles, and is beaten with a stick or kneaded with the hands, washed and strained through coconut fibre or a bunch of sago leaves (Plate 44). In some areas the pith may be trampled in a bag laid on a platform of sago leaf midribs. The water and suspended starch runs down into a settling tank made of sago palm bark, sago leaf midrib or spathe or canoe of sago bark fitted with a second strainer. The water is tipped off, or bailed out of the tank with a coconut shell or sago spathe, and the starch pounded into cylindrical lumps, which are glazed on an open fire and then wrapped in nipa or sago palm leaves for carrying in a sago leaf petiole.

In many areas where large stands of sago are close at hand, the flour is prepared each day but, if distances are great between sago areas and the village, large quantities are made and stored in the houses over the fire. In some cases the raw pith may be carried back to the village and then washed in a trough of coconut leaves set up near a river (Hogbin 1938-9).

Plate 44
A woman making sago on the banks of the Mubi River, near Lake Kutubu

Barrau (1959) records average yields of sago as 110-136 kg per trunk from near-flowering plants and up to 400 kg from plants that are naturally sterile. In swamp forest groves in New Guinea there are about 60 palms per ha per year worth felling; these produce 7000-9000 kg of crude starch with a water content of 35-40 per cent. The daily ration, where sago is the staple food, is about 1 kg, representing about 1.6 kg raw sago, 1700-4000 calories.

The sago may be baked in leaves or pottery bowls, roasted on an open fire or simply mixed into a mush or porridge by pouring boiling water over it. It is a poor food on its own and is usually eaten with fish or with beans, bananas, coconut and greens. Chemical analyses indicate it comprises mainly starch and water; protein content is about 0.2 g, calcium 10 mg, phosphorus 12.5 mg, iron 1.5 mg per 100 g of edible portion. Vitamins and fats are absent (Barrau 1959).

The other plant foods eaten with sago may be traded in from adjacent areas or grown nearby. In the Waropen area where the people live in houses built on piles in the mangrove forests, sago and fish are considered to be a complete diet. The almost aquatic *Ipomoea reptans* is gathered from river banks, and condiments *Capsicum frutescens* and *Amaranthus* spp. are grown in 'hanging gardens' built on piles near the house; all other foods, taro, bananas, coconuts, sugarcane, malay apples (*Syzygium malaccense*) and wild nutmeg (*Myristica* sp.) fruit are traded in (Held 1957; Oomen and Malcolm 1958). In other areas, islands or areas of higher alluvial ground are available for growing crops. Yams, taro and bananas are the main crops in these small, mixed gardens; coconuts and breadfruit are usually present, with *Amaranthus* spp. and *Hibiscus manihot* as green vegetables. The gardening system practised is generally a long-fallow rotational system involving slash and burn clearance, little if any tillage and abandonment after 1-2 years' use.

An important supplementary food in the mangrove swamp is the sprouting fruits of *Bruguiera* spp. These are harvested, boiled in sea water to soften the skin and then peeled, sliced and soaked in sea water until soft. They are then placed in a woven bag, the water squeezed out and the fruit kneaded into a soft mash. This is rolled into balls and eaten with the fingers. An aquatic, *Enhalus* sp., and a seaweed are also gathered as supplements in some areas.

In many areas of fresh water swampland *Cyrotosperma chamissonis* and *Xanthosoma sagittifolium* tubers are used; so also are bamboo shoots, the heart 'cabbage' of the sago palm and of *Arenga* sp. and *Caryota* sp. The young fruits of sago and the soft endosperm of immature seeds of nipa (*Nypa fruticans*) are eaten also. Forest resources (on higher ground) provide edible leaves of *Ficus* spp., *Euodia* sp., *Gnetum gnemon*, *Cassia* sp. and *Albizia* sp., edible petioles of *Oncosperma* sp. and ferns and fungi. Fruits and nuts of *Terminalia catappa, Myristica* sp., *Aglaia* sp., *Garcinia* sp. and seeds of *Gnetum gnemon* and the wild breadfruit, *Artocarpus* sp., may be gathered also (Conroy and Bridgland 1950). The food values of some of these species are given in Tables 3.2 and 3.3. In some areas the leaves and/or fruits of recently introduced

Table 3.2 Chemical composition of staple crops

Composition per 100 g edible raw portion	*Colocasia esculenta*	*Dioscorea* spp.	*Ipomoea batatas*	*Manihot esculenta*	*Musa* spp.	*Metroxylon* spp.
Moisture	54-73	55-75	59-76	60-67	62-76	17-40
Calories	100-165	95-161	90-150	130-150	85-142	285-362
Protein (g)	1.4-2.0	1.0-2.5	0.5-3.4	0.5-1.2	1.0-1.3	0.1-0.5
Fat (g)	0.2-0.5	0.05-0.2	0.1-0.4	0.3	0.1-0.5	0.0-0.3
Carbohydrates (g)	24.2-39.0	15.0-38.7	22.0-34.0	30.0-35.0	20.0-34.6	71-86.9
Fibre (g)	0.7-3.0	0.5-1.5	0.5-1.0	0.9-1.3	0.4-1.6	0.2-0.3
Calcium (mg)	15-39	10-23	15-40	10-33	7-8	0-10
Iron (mg)	1.0-1.1	0.5-1.2	0.6-1.0	0.5-0.9	0.4-0.8	0.0-1.5
Pro-Vitamin A (I.U.)	5-30	0-20	0-4000	—	200-400	—
β-Cartoene (μg)	0-3	0-12	0-2400	—	120-190	—
Thiamin (mg)	0.04-0.27	0.09-0.40	0.07-0.15	0.02-0.06	0.04-0.06	0
Riboflavin (mg)	0.03-0.04	0.03	0.04-0.05	0.03	0.04-0.08	—
Niacin (mg)	0.20-0.90	0.19-0.50	0.24-0.70	0.60	0.60-0.70	—
Ascorbic acid (mg)	5.0-10.0	0.0-15.0	23.0-70.0	30.0-36.0	10.0-31.3	0

References: Hipsley and Clements 1950, Massal and Barrau 1956, Peters 1958, Purseglove 1968, 1972.
Note: In this and following tables — means that no information is available and 0 means zero.

species including cucumber (*Cucumis sativus*), watermelon (*Citrullus vulgaris*), papaya (*Carica papaya*), tomatoes (*Lycopersicon lycopersicum*), beans (*Phaseolus aureus, P. vulgaris*), pineapple (*Ananas comosus*) and *Citrus* spp. are used (Luyken and Luyken-Koning 1955).

In very many other areas sago is an important supplementary food or is used as an occasional supplement, produced locally or traded in for feasts (Lea 1965; McAlpine 1968; Fagan and McAlpine 1972; Williams 1930; Hatanaka and Bragge 1973; Floyd 1954; Hogbin 1938-9; Serpenti 1965; Sorenson and Gajdusek 1969).

Mixed crop swamp cultivation (Frederik-Hendrik Island). An extremely complex system of swamp cultivation has been developed on Frederik-Hendrik Island, off the south coast of Irian Jaya. Artificial islands usually 2-3 m but up to 8 m wide are built in the fresh water swamps by cutting the natural reed growth, plastering it with clay and then building up the island height with alternate layers of drift grass and clay. A 2-3 m wide ditch is maintained around the island to attempt to keep the water table at the desired level. Yams are the main crop grown (sixteen varieties of *Dioscorea alata* and six of *D. esculenta* are recorded) but taro (twenty-three varieties) and sweet potato (fifteen varieties) and the recently introduced cassava (*Manihot esculenta*, four varieties) are also important. Cultivation techniques employed are complex and the growing crops are fertilised from time to time with drift grass and organic mud. As the taro varieties grown can tolerate a high water level they are planted usually on the larger, lower islands; after harvesting the ground may be used for cassava, but is usually left to lie fallow for 1-2 years before being replanted with taro. The higher, drier, but smaller islands are used for yams, sweet potato and cassava, with bananas and sugarcane interplanted. Yam-cassava-yam, sweet potato-cassava-yam or yam-fallow-yam rotations are common.

Elsewhere low islands are planted with sago palms, an important

Table 3.3 Chemical composition of some supplementary crops

Species	Moisture	Calories	Protein	Fat	Carbo-hydrates	Fibre	Calcium	Iron
			g	g	g	g	mg	mg
Fruit and nuts								
Ananas comosus	87	47	0.5	0.2	12.2	0.5	18	0.5
Annona muricata	80	71	0.8	0.4	18.0	1.0	20	0.5
Artocarpus altilis — flesh	69-80	55-110	0.8-4.3	0.3-1.5	12.0-26.0	1.2-5.4	2-35	0.5-1.2
— seeds	52	150	6.0	0.5	30.0	4.0	60	—
Bruguiera sp.	72	75	2.5	0.3	16.0	4.0	—	—
Canarium spp.	9	644	14.2	68.5	5.5	3.2	119	2.6
Carica papaya	87-89	39-45	0.5-0.6	0.1	10.1-12.1	0.9	20-24	0.4
Citrus sp.	83-90	41-59	0.4-0.9	0.2-1.0	8.3-15.3	0.6-1.0	18-40	0.4-0.8
Coix lachryma-jobi	12	140	15.0	7.0	60.0	3.0	70	—
Ficus spp. — fruit	80	70	1.3	0.2	17.8	—	30	—
Garcinia spp.	90	23	0.6	+	6.8	1.8	5	—
Gnetum gnemon — seeds	81	41	3.3	+	9.0	5.7	53	—
Inocarpus edulis	43	240	4.5	4.5	40.0	2.5	—	—
Macaranga spp.	9	601	18.9	60.4	7.4	2.2	474	—
Mangifera spp.	82	65	0.7	0.2	17.0	0.8	11	0.4
Myristica spp.	90	20	0.4	+	7.3	1.9	58	—
Pandanus spp. — nut	9	683	11.9	66.0	22.0	6.1	419	—
Psidium guajava	81	69	1.0	0.4	17.3	6.2	18	0.9
Saccharum officinarum	82	58	0.4	+	14	10-15	10	+
Syzygium malaccense	85-92	13-51	0.4-0.6	0.1	3.6-13.7	0.7-1.5	0-10	0.4-0.5
Trichosanthes — flesh	88	44	1.5	0	99.0	—	20	—
— seeds	13	594	26.0	50.0	10.0	—	60	—
Vegetables								
Allium spp.	88	43	1.8	0.2	9.4	1.2	80	1.0
Alsophila spp.	81	43	5.5	—	6.3	—	—	—
Amaranthus spp.	86-90	6-50	3.0-5.1	0.5	4.2-7.0	1.3-1.5	267-300	3.9-30.0
Archontophoenix sp.	92	19	1.9	0	3.2	1.9	550	—
Athyrium esculentum — fronds	92	20	2.4	0.3	2.0	1.0	20	8.0
Bambusa spp.	90-91	27-35	2.6	0.3	5.2	0.7	13	0.5
Brassica spp.	88	35	1.9	0.1	7.0	1.0	30	0.2
Brassicacea (?*Rorippa* sp.)	86	40	4.4	—	6.8	—	—	—
Capsicum frutescens	72	62	4.8	2.2	9.0	1.4	65	2.3
Colocasia esculenta — leaves and petioles	85-93	19-57	0.5-5.0	0.0-2.5	3.9-7.4	1.5-2.0	150-400	1.0
Cucumis sativus	96	13	0.7	0.2	2.0	0.5	15	0.2
Cucurbita maximal pepo								
— leaves	89	35	3.8	0.4	4.0	1.0	100	31.0
— fruit	92	25	1.5	0.1	5.0	1.3	10	0.7
Cyrtosperma spp.	63	140	0.8	0.5	33.0	1.5	—	—
Dolichos lablab — seeds	12-13	338-340	22.0-28.2	1.1-1.5	60.0-66.7	6.0-7.7	88	3.5
Dryopteris spp. — fronds	78-85	52-59	3.4-5.5	0.0-0.2	9.9	—	49-370	—
Ficus dammaropsis	86	38	22.0	—	7.4	—	200	—
Gnetum gnemon — leaves	76	1	6.4	—	—	—	240	—
Hibiscus manihot	83	31-47	5.6-5.7	0.3	6.8-8.6	1.8	580	4.0
Ipomoea aquatica	90	30	3.9-4.0	0.5-0.6	3.0-4.4	1.0	50-71	3.2-4.5
Ipomoea batatas — leaves	85	40-47	3.5	0.5	8.0-9.5	1.5	70	8.0
Metroxylon spp. — 'cabbage'	92	—	1.6	0	1.3	—	7	—
Musa spp. — flowers	90	26	1.0	0.2	5.0	2.0	30	—
Phaseolus lunatus — seeds	12-13	338	20.7-25.9	1.1-1.3	57.3-67.6	4.3-5.7	90	6.0
Phaseolus vulgaris	89	35	2.4	0.2	7.6	1.5	57	0.8
Psophocarpus tetragonolobus — seeds	14	399	29.8-37.3	15.0-19.1	28.0-42.0	5.0-6.7	—	—
Pteris moluccana — flour	15	272	1.0	0.0	67.0	10.0	—	—
Saccharum edule	89	34-38	3.8-4.1	0.0	6.9-7.6	0.7	10	0.4-21.0
Setaria palmifolia	92	25-39	0.5-2.6	0.2	5.2-6.8	1.1-1.2	7-21	0.2
Zea mays	74	92	3.4	1.2	20.7	1.0	5	0.6

Species	Pro-Vitamin A I.U.	β-Carotene μg	Thiamin mg	Riboflavin mg	Niacin mg	Ascorbic acid mg
Fruit and nuts						
Ananas comosus	90	54	0.08	0.03	0.2	40
Annona muricata	20	12	0.06	0.06	1.0	19
Artocarpus altilis — flesh	19-40	11-24	0.08-0.15	0.03-0.06	0.8-0.9	23-72
Canarium spp.	45	27	0.95	0.12	0.4	+
Carica papaya	1000	600	0.03-0.15	0.04	0.4	29-64
Citrus spp.	0-420	0-250	0.02-0.08	+-0.03	0.1-0.2	31-50
Coix lachryma-jobi	—	—	0.30	—	—	—
Ficus spp. — fruit	—	—	0.06	—	—	2
Garcinia spp.	—	—	—	—	—	5.3
Gnetum gnemon — seeds	—	—	—	—	—	148
Macaranga spp.	—	—	—	0.08	—	—
Mangifera spp.	1900	1140	0.05	0.06	0.6	48
Myristica spp.	—	—	—	—	—	6.0
Pandanus spp.	—	—	—	0.36	—	—
Psidium guajava	180	110	0.03	0.04	1.2	160
Saccharum officinarum	—	—	+	—	—	+
Syzygium malaccense	+	+	0.03	0.03	0.3	3-15
Trichosanthes sp. — flesh	—	—	—	—	—	52
Vegetables						
Allium spp.	50	30	0.06	0.04	0.5	18
Alsophila spp.	—	—	0.15	—	—	25.5
Amaranthus spp.	6090	3600	0.01-0.15	0.16	1.7	28-80
Archontophoenix sp.	—	—	—	—	—	14
Athyrium esculentum	—	—	0.0	—	—	15
Bambusa spp.	20	12	0.15	0.07	0.6	4
Brassica spp.	—	—	0.06	—	—	50
Brassicacea (?*Rorippa* sp.)	—	—	0.15	—	—	56
Capsicum frutescens	7010	1200	0.31	0.25	1.8	69
Colocasia esculenta — leaves	—	—	0.10-0.20	—	—	16-214
Cucumis sativus	—	—	0.03	—	—	15
Cucurbita maxima/pepo	—	—	0.06-0.16	—	—	5-50
Dolichos lablab — seeds	—	—	0.10-0.60	0.14-0.20	0.9-2.3	21
Dryopteris spp.	—	—	0.10-0.15	—	—	47-56
Ficus dammaropsis	—	—	0.10	—	—	5.1
Gnetum gnemon — leaves	—	—	0.10	—	—	390
Hibiscus manihot	13000	—	0.15	—	—	118
Ipomoea aquatica	4825	3000	0.09-0.12	0.24	1.3	49-60
Ipomoea batatas — leaves	6000	3600	0.10	0.20	0.9	25
Metroxylon spp. — 'cabbage'	—	—	—	—	—	15.7
Musa spp. — flowers	—	—	—	—	—	13
Phaseolus lunatus — seeds	+	+	0.22-0.50	0.14	1.29-1.5	+
Phaseolus vulgaris	400	240	0.08	0.12	0.5	17
Psophocarpus tetragonolobus — seeds	—	—	0.08	—	—	37
Saccharum edule	—	—	—	—	—	21
Setaria palmifolia	—	—	—	—	—	8
Zea mays	350	210	0.15	0.09	1.7	14

References: Hamilton 1955, Hipsley & Clements 1950, Massal & Barrau 1956, Oomen and Malcolm 1958, Peters 1958, Powell 1974b.

+ = traces present

supplementary crop, and higher islands and dwelling islands with coconuts, bananas and introduced fruit trees (papaya, soursop, citrus). In most parts of the island the roots of the fern *Nephrolepis biserrata* are used as a supplementary food during the wet season and as a famine food when required. The roots are gathered and, if required immediately, are simply dried, peeled and grilled over the fire. Usually, however, they are soaked in water for 1-3 months, then sun-dried, scraped and pounded into flour in a canoe-shaped trough. The flour is sifted through a finely plaited mat and stored in plaited bags until needed. It is then mixed with a little water, rolled into balls and wrapped in leaves to be cooked on hot stones or grilled over the fire (Serpenti 1965).

The lowland shifting agriculturalists

Throughout the non-swampy lowland areas of New Guinea a system of swidden gardening with long-term bush fallow is practised. To clear an area of forest or savanna the undergrowth is first slashed, the small trees felled and the larger trees cut or simply ring-barked or pollarded. The debris is then stacked around the tree stumps and left to dry for several weeks before being burnt. Saplings and poles are sorted out for use as boundary markers and for fencing and firewood. Once the debris is burnt the women usually clear the ground further while the men construct the boundary fence. Individual gardens within the whole area are then allotted and marked out with horizontally placed logs and saplings. In coastal areas coral boulders may be cleared away or used as boundary markers.

Planting of the crops is usually undertaken by both men and women using their bare hands and/or a pointed wooden digging stick. Little tillage of the soil is undertaken. The main crops grown throughout the lowlands are yams, taro and bananas, and occasionally sweet potato and cassava. In mixed crop gardens a fairly definite planting pattern is often followed, with the staple crops planted throughout the garden, green vegetables, condiments and tobacco in ashy areas near tree stumps and fruit trees around the margin. Even in single crop gardens a few fruit trees and useful shrubs are interplanted.

The garden is weeded several times before the crops are harvested and then it is left to fallow. Fruit, nuts and other useful products are harvested from the young regrowth but after 2-3 years the garden is abandoned completely to fallow for twenty years or longer. Garden magic may be practised at all stages of the gardening cycle but is emphasised at planting times and during subsequent cultivation and harvesting; it is particularly important for crops being grown for ceremonial purposes such as the giant yams of the Trobriands and Sepik areas (Malinowski 1935; Kaberry 1941; Lea 1966).

Staple crops in grassland and savanna areas: yams, bananas and cassava. Lowland grassland and savanna areas are restricted mainly to the southern parts of Irian Jaya and Papua New Guinea and the broad valleys of the north coast of Papua New Guinea. Rainfall in these areas may be up to 2540 mm a year but it usually shows marked seasonality. Population densities are relatively low (2-4 persons per km²) but may be as high as 16-30 persons per km² in local coastal areas.

Yams (*Dioscorea* spp.) are the staple crop in some of these seasonally dry areas: they are recorded as such from parts of the Trans-Fly, the coast further east, the Trobriand Islands and parts of the north coast and inland Sepik plain (Williams 1936; Hipsley and Kirk 1965; Malinowski 1935; Austen 1945; Kaberry 1941; Lea 1965; McAlpine 1968; Fagan and McAlpine 1972). *D. alata* and *D. esculenta* are the two most widely grown species (with very many named cultivars) but *D. bulbifera, D. pentaphylla, D. hispida* and *D. nummularia* are also present and often grown in small quantities. *D. pentaphylla, D. hispida* and an unidentified *D.* sp. also grow wild; they may be harvested on hunting trips and are important in 'famine' times.

Yams grow best on deep, fertile, well drained soils and they are usually the first crop planted in any rotation. They are a seasonal crop. The planting materials are whole small yams (Chowning pers. comm.), sections of larger yams that have been stored in the house until they sprout, or aerial bulbils if produced (as on *D. bulbifera, D. alata*); these are planted into a mound or raised bed of light friable soil.

In the Sepik, where tubers are grown for ceremonial purposes, large holes are dug and lined with sticks to allow for the full development of the tuber, and long stakes are pushed into the ground to support the growth of the vine. After 9-12 months' growth the mature tubers are carefully harvested, sorted and stored in specially built cool and airy yam houses. If kept in dry, dark conditions the yams will last for six months or more.

Lea (1966) gives some yield data for various species of yams grown for ceremonial and for everyday consumption by the Abelam. The average yield in ordinary mixed crop gardens, where the yams were closely spaced, was 9 kg per plant or 21 metric tons per ha, while yields of 18-30 kg per plant were recorded for ceremonial yams (widely spaced) with an average of 20 kg per plant and 16 metric tons per ha. At Yenigo 29 metric tons of ceremonial yams were produced per year, compared with 289 metric tons of non-ceremonial yams.

Yams can be roasted on the fire, boiled or baked in stone ovens, or grated raw, mixed with coconut or green vegetables and baked in banana leaves as a pudding (Barrau 1958). Chemical analyses (Table 3.2) indicate that yams provide a good source of calories and appreciable amounts of minerals and some vitamins; protein content is low.

In many areas where yams are the staple, bananas and taro, and occasionally also cassava, are important supplementary crops. Other cultivated crops include the edible pitpit *Saccharum edule* and the beans *Phaseolus lunatus* and *Psophocarpus tetragonolobus*, the green vegetables *Hibiscus manihot, Amaranthus* spp. and occasionally *Setaria palmifolia* and *Oenanthe javanica*. Sugarcane is always important. Coconuts and breadfruit trees are always planted in village areas, papaya and *Ficus* spp. often, and other fruit and nut trees occasionally. Palm shoots, gingers, leaves of *Gnetum gnemon, Ficus* spp., *Convolvulus* sp. and wild yams, breadfruit seeds and legumes are gathered from the forest. Many of these species are important in providing extra protein, minerals and vitamins in the diet (Tables 3.2 and 3.3). In coastal areas near Port Moresby and in the Markham valley bananas (*Musa* spp.) are the staple crop. Very many varieties, belonging to both the Eumusa and Australimusa sections of the genus, are grown. Bananas can be grown on a wide range of soil types but do best in moist, fertile ground. They are planted in mixed crop gardens, near houses or in semi-permanent stands. Suckers are used as planting material and few special techniques are employed in growing them. The plant takes 12-18 months to mature. The ripening fruit may be wrapped in dry leaves to protect it from birds and bats, or breadfruit latex may be spread around (Hogbin 1938-9). Frequently the unripe fruit is harvested and in some areas is buried in mud until ripe (Barrau 1958).

Yields of bananas growing in subsistence garden context have not been studied in detail. Massal and Barrau (1956) suggest that bananas growing sparsely in fallow areas could yield about 1.3 metric tons of fruit per ha, while those grown in plantations or semi-permanent groves (as the staple) yield about 10-20 metric tons per ha, representing some 7-14 million calories per ha.

Most varieties grown are cooking bananas and they are baked, boiled or roasted in their skins on the open fire before being eaten. Some types may be eaten raw when ripe and these are occasionally dried (Oomen and Malcolm 1958). Analyses of chemical composition of unripe bananas (cooking bananas) show they are similar to other starchy crops in protein, carbohydrate and fat content. Ripe bananas contain sugar rather than starch and hence are more easily digested. All bananas are a good source of vitamin A and C but are poor in vitamin B and calcium.

Cassava, *Manihot esculenta,* is the other main tuber crop grown in dry grassland and savanna areas. Although a post-European contact introduction, it has spread rapidly and in some areas replaces earlier staples (Williams 1936; Lea 1970). It is tolerant of a wide range of environmental conditions and will grow in drier and poorer soils than either yams or bananas. It is planted from stem cuttings at any time of the year and matures in 6-12 months; it is relatively high yielding, producing 15-20 metric tons of tubers per ha, representing some 16-22 million calories. Nutritionally, cassava is a very poor food. While calcium and vitamin C values are relatively high (Table 3.2), the protein content is very low and some essential amino-acids are absent. Moreover, since about 25 per cent of gross weight is lost in preparation, the actual energy value of 100 g of fresh cassava is only 90-95 calories. Some varieties grown contain a hydrocyanic glucoside poison and these must be peeled, grated and washed well before being cooked.

Staple crops in rain forest areas: taro and sweet potato. Rain forest dominates the coastal plains, foothills and mountains where the rainfall is higher and more evenly distributed throughout the year. In most areas rainfall exceeds 2540 mm per year and in some parts (along the southern edge of the central cordillera and southern coast of New Britain) is over 5080 mm per year. Population densities vary up to 8-16 persons per km².

Taro (*Colocasia esculenta*) is the staple crop grown in rain forest on the north coast of New Guinea and on all the major islands (Hogbin 1938-9). 1951; Wedgwood 1934; Williams 1930; Todd 1934; Panoff 1972). It has been recorded as the staple also from parts of Biak Island, the Vogelkop peninsula, Star Mountains and upper Fly and Strickland River areas (Oomen and Malcolm 1958; Reynders 1962; Barth 1971; Hatanaka and Bragge 1973). Bananas, sweet potato and yams are generally associated as supplements; so also are *Xanthosoma sagittifolium* and cassava (Plate 45).

Plate 45
Taro grown with sweet potato on drained swampland at Pureni, Tari Subdistrict

Taro requires moist, shady conditions for adequate growth and is often planted in swampy conditions. In some cases it is irrigated (Lea 1972). It is propagated by planting the top of the corm and attached shoot, axillary buds or adventitious runners. Cultivation techniques are simple, the taro shoot being placed in the bottom of a 10-15 cm diameter hole made with a pointed digging stick or wooden club, and left uncovered. It can be planted at any time of the year and the corms are ready to be harvested after 5-12 months' growth, depending on soil fertility and the variety grown. Very many varieties of taro are distinguished on the basis of leaf and petiole colour, corm flesh colour and time of maturity. Yields vary considerably; an average of 8-15 metric tons of corms per ha, or about 6-13 million calories, is recorded by Massal and Barrau (1956). Large taros grown for feasts may weigh up to 45 kg but usually they are harvested when much smaller (Hogbin 1938-9).

Taro is usually eaten immediately after being harvested but it may be stored for a short period without any special treatment (Chowning, pers. comm.) or laid in a grass-filled hole and covered with further grass and soil (Williamson 1912). It is cooked by the usual methods: roasting, baking and boiling. It may be sliced up and added to green vegetables, meat or fish and coconut cream before being baked in a stone oven, or grated raw, mixed with various nuts (*Canarium* spp., *Barringtonia* sp.) and baked in banana leaves as a pudding. Taro is nutritionally similar to the other root crops. It is a good source of calories and the vitamin B group and vitamin C and mineral salts are adequately represented; the protein and fat content is low.

Sweet potato (*Ipomoea batatas*) is an important supplementary crop in rain forest areas of the Huon peninsula and further east, and has replaced taro as the staple in some areas of Bougainville in recent years following outbreaks of the taro fungus disease *Phytophthora colocasiae* (McAlpine 1967). Sweet potato is easy to grow and tolerant of a wide range of climatic and edaphic conditions. It is often grown as a second crop following yams or taro in a mixed crop garden rotation. It is grown from cuttings, usually planted directly into the ground; it does not tolerate waterlogging, however, and in some cases raised beds or high mounds are used for its cultivation. It matures in coastal areas in 3-6 months, depending on variety, and yields some 8-15 metric tons of tubers per ha according to Massal and Barrau (1956); in experimental plots in lowland Irian Jaya yields were less than 8 metric tons per ha (Ruinard 1967). The tubers are harvested when needed, as they do not store well, and are cooked in the usual manner. The nutritional value of the sweet potato is indicated in Table 3.2. Wide variations occur in protein contents and in mineral constituents and these may be affected by soil conditions (Kimber, pers. comm.; Yen 1974). The proteins are considered to be of relatively high quality and adequate quantities of minerals and vitamins are usually present; vitamin A content is particularly high in yellow and orange-fleshed varieties (Massal and Barrau 1956). Sweet potato will be considered further below as the staple of the highlanders.

Fruit and nut crops. Fruit and nut trees are extremely important in coastal and lowland areas as supplementary crops. Coconuts (*Cocos nucifera*) are ubiquitous; they occur naturally on the strandline and are always planted in villages and often also in groves nearby. The meat of the nut can be eaten at all stages of development and the mature meat, grated and mixed with water and then strained through coconut fibre (or squeezed by hand) provides a cream that can be guaranteed to enhance the flavour of any vegetable or fish dish being prepared. Raw grated coconut is also mixed with starchy foods being prepared as porridges or puddings. The milk of the young, green coconut is a refreshing drink. At certain times of the year coconuts may become the staple plant food, as on the Rai coast (Miklouho-Maclay 1885). They are a valuable food mainly because of their high fat content (Table 3.4). As the nut matures and the albumen develops the moisture content decreases while the calorific value and the fat, carbohydrate and protein contents rise. The mature nut is a good source of iron and phosphorus (98-170 mg per

Table 3.4 Chemical composition of coconuts (*Cocos nucifera*)

Composition per 100 g edible portion	Mature 'meat'	Immature 'flesh'	Mature nut milk	Mature nut water	Immature nut water
Moisture	36-50	70-92.8	59	94-95	95
Calories	351-500	30-180	311-346	13-32	11-18
Protein (g)	4.0-4.7	0.7-4.0	2.5-4.3	0.3-0.5	0.1-0.2
Fat (g)	33.0-53.4	1.0-15.0	34.0-35.0	0.3-1.5	0.1-0.4
Carbohydrates (g)	5.1-15.0	4.0-10.0	5.0-6.0	2.0-5.0	2.6-4.0
Fibre (g)	3.0-3.6	0.5-3.0	0.0	0.0	—
Calcium (mg)	9-25	6-8	11-15	20-30	30
Iron (mg)	1.7-2.2	0.8-1.3	1.0-2.3	0	—
Pro-Vitamin A (I.U.)	0	10	0	—	—
β-Carotene (μg)	0	0-6	0	—	—
Thiamin (mg)	0.04-0.15	0.05	0.02-0.03	0	0.002
Riboflavin (mg)	0.01-0.03	0.02-0.03	0.01	0	0.08
Niacin (mg)	0.6	0.3-0.5	0.3-0.9	0	0.1-0.2
Ascorbic acid (mg)	2.0-38.0	4.0-5.3	0.0-2.8	2-24	1.4

References: Hipsley and Clements 1950, Purseglove 1972.

100 g edible portion) but is poor in other minerals and most vitamins (Massal and Barrau 1956). The apical bud or 'cabbage' of the palm is eaten occasionally and in a few areas a toddy is prepared by tapping the sap of the unopened flower spathe; by percentage it comprises 84.4 water, 0.1 protein, 0.1 fat, 15.1 carbohydrate and 0.3 minerals (Purseglove 1972). Coconut palm products provide the people with almost all their raw material needs; these will be discussed later.

Breadfruit, *Artocarpus altilis*, is also an extremely important fruit tree of widespread occurrence in the New Guinea lowlands. Wild plants are usually individually owned and tended; the surrounding forest growth is cleared away from the base of the tree and a fence may be erected. Seedlings are frequently transplanted to village areas and older plants may be propagated by planting root cuttings or suckers. Both the fruit pulp and the seeds are eaten in some varieties while in others the seeds only are edible, the flesh being tough and stringy (Oomen and Malcolm 1958; Floyd 1954). Breadfruit is highly valued in some areas and may even replace the staple in its season. The fruit is roasted or baked whole or cut up and boiled; it can also be sun-dried and stored for later use (Hogbin 1938-9). Seeds are usually roasted or boiled. The chemical composition of the fruit is given in Table 3.3. While it provides a satisfactory source of calories, minerals and vitamins, its protein and fat contents are low; the seeds, with a higher protein content, are nutritionally more valuable.

A number of other fruit and nut trees are of importance in New Guinea coastal and lowland areas but have received little attention from botanists, agriculturalists or nutritionists; their value as supplementary foods has almost certainly been underrated (Table 3.3). *Canarium* spp. are widespread in coastal and island New Guinea and form an important seasonal supplement in some areas (Hogbin 1938-9). Individually owned trees are tended in the forest and seedlings are often transplanted into the villages. Two species are used: *Canarium indicum*, (including *C. commune* and *C. mehenbethane*) and *C. salomonense*; they are also important in the

Solomon Islands (Whitmore 1966a). The nuts are broken open with stones, the kernels eaten raw or crushed and added to sago and starchy root puddings. If dried over the fire they can be stored for a long time. *Terminalia catappa* and *T. kaernbachii* are frequently planted (Coode 1969a) and other species with edible kernels (*T. copelandii, T. impediens*) are harvested from the forest. The kernel is difficult to extract but provides an excellent side-dish, eaten at feasts or as a luxury with roasted tubers. *T. solomonensis*, common in villages in the Solomon Islands and also in east New Guinea, has an edible fruit (Whitmore 1966a). Nuts of the Tahitian chestnut, *Inocarpus edulis*, are gathered widely also; they need to be cooked to be palatable.

Several species of *Syzygium* have edible fruits and are planted (*S. malaccense, S. aqueum*) or tended within the forest. *Pometia pinnata*, widely used in the Solomon Islands (Whitmore 1966a), is a favourite fruit of the Sengseng of New Britain and eaten elsewhere also, while *Parartocarpus venenosa* is another important fruit tree there, often cultivated (Chowning, pers. comm.). A *Barringtonia* sp. is recorded by Miklouho-Maclay (1885) as a forest plant with edible nut on the Rai coast and is probably more widespread; *B. edulis* is a common village fruit tree in the Solomon Islands (Whitmore 1966a). *Burckella* sp. fruit is also eaten in New Guinea (Massal and Barrau 1956) and the Solomon Islands (*B. obovata*, Whitmore 1966a). Miklouho-Maclay (1885) mentions two species of *Madhuca* (*Bassia*) with edible fruit that were harvested from forest trees. He also mentions the use of the fruit of *Pangium edule*; it was hung in baskets and allowed to ferment to produce an acidic, strong-smelling sauce, which was mixed with other foods and considered a delicacy. The fruit of various species of *Diospyros* and *Macaranga* are also edible.

The seeds of the red or orange fruits of *Gnetum gnemon* are roasted in some areas, but more often the leaves are used as a vegetable. A number of coastal *Pandanus* spp. bear edible fruits, and *Morinda citrifolia, Spondias dulcis, Garcinia* spp. and *Mangifera* sp. are also harvested. Many other species used remain unidentified at present. Other mangoes, pineapples, papaya, guava, *Citrus* spp. and *Annona* spp. have spread fairly rapidly since their introduction by Europeans (Barrau 1958).

Green vegetables. Widely cultivated as green vegetables in coastal and lowland areas are 'Aibika', *Hibiscus manihot,* and spinach-like forbs, *Amaranthus* spp. *Hibiscus manihot*, a low, much-branched shrub, is planted by cuttings in mixed crop gardens and near houses and harvested also from young regrowth in abandoned gardens and forest margins. Very many varieties are grown; they differ in leaf size and shape and stem colour. The leaves and young shoots are cooked in stone ovens, boiled or steamed in bamboo containers or roasted in bark. They are often eaten mixed with the young leaves and petioles of taro. The plant's main nutritive value lies in its high protein to calorie ratio; it also has high mineral and vitamin content (Table 3.3). *Amaranthus* spp. (*A. tricolor, A. hybridus, A. viridis*) are grown from seed broadcast in the mixed gardens. The whole young plant is gathered after 1-2 months' growth, steamed in bark or bamboo or cooked in an earth oven with other vegetables, and usually eaten with meat or fish. Other plants are left to produce seed, which is dried and stored in the house until needed. These spinaches have a relatively high protein content (3.5-5.1 per cent) and more than adequate quantities of iron and calcium and vitamins A and C (Table 3.3).

The young leaves and petioles of some taros (*Colocasia esculenta, Xanthosoma sagittifolium*) are commonly eaten as vegetables. Some varieties require special preparation and prolonged cooking as they contain appreciable amounts of oxalate crystals. An edible pitpit, *Saccharum edule*, is an important seasonal food in some areas. Cuttings or shoots 0.5-1.0 m long are planted, two or three in a group, in mixed crop gardens. Varieties differ in stem colour, leaf form and the position of the inflorescence. They take 6-9 months to mature, the edible part being the young inflorescence. In almost all varieties the inflorescence does not mature; if left unharvested it rots within its leafy sheath. The pitpit may be roasted in its leaf sheath directly over the open fire or cooked in coconut cream with other vegetables in bark or bamboo containers. It has a relatively high protein content, 3.8-4.1 per cent. *Setaria palmifolia*, another edible grass, is important in some lowland areas, for example New Britain (Chowning, pers. comm.), but is much more widely grown in the highlands. The soft inner shoot is eaten raw or cooked, and has some nutritive value (Table 3.3). It is planted by shoot cuttings and takes 3-5 months to mature. The many cultivated varieties, differing in shoot colour and size and growth form, do not flower. Wild forms, some of which do flower, are gathered from river banks and forest areas.

Hyacinth beans (*Dolichos lablab*) and winged beans (*Psophocarpus tetragonolobus*) are grown in some lowland areas, for example Fergusson Island (Chowning, pers. comm.), as seasonal crops together with the recently introduced french beans, *Phaseolus vulgaris*. They form a valuable source of protein, providing almost all essential amino acids. *Oenanthe javanica* and *Commelina* spp. may also be cultivated occasionally in coastal and lowland areas but are more important in the highlands. The edible shoots of a number of bamboo species are harvested and leaves and shoot tips of sweet potato, pumpkin and wild Cucurbitaceae are frequently used as vegetables. Leaves and fruits of figs (*Ficus* spp.) are commonly eaten. Ferns are important, especially for feasts; species of *Asplenium, Athyrium, Ctenitis, Cyathea, Diplazium* and *Dryopteris* are recorded as being used, together with *Cyclosorus truncatus* and species of *Dennstaedtia*. As 'famine' foods, wild yams are very important (*Dioscorea* spp.); others used are cycads (*Cycas* spp.) and the pith of a palm, *Archontophoenix* sp., the latter being processed into flour or simply roasted over the fire before being eaten (Chowning, pers. comm.).

Sugarcane. *Saccharum officinarum*, the sugarcane, is grown throughout lowland and highland areas of New Guinea. It is not considered an important food crop but is used for refreshment, being chewed at any time during the day. Many different varieties are grown, recognised and named on the basis of their cane colour, growth form, sweetness and fibre content. Cuttings of sugarcane are planted at any time of the year, in gardens and sometimes near rivers where the ground is moist and rich. The canes may be harvested after a minimum of twelve months' growth but, if required for ceremonial purposes, they are tied together or staked, wrapped in leaves and left to grow for 2-3 years or longer. Sugarcane is considered to be a good food for sick people and is very important in ritual medicine.

The highland margins

Areas marginal to the large highland intermontane valleys have received relatively little attention. The population density is generally fairly low and the people may be more or less nomadic, gathering sago from lower elevations at certain times of the year and cultivating small gardens in forest at other times (Hatanaka and Bragge 1973; Barth 1971; Reynders 1962). Shifting swidden agriculture is practised with single cropping followed by a long fallow of 15-30 years or more. Taro is the main crop grown in most areas, with sweet potatoes and bananas as important supplements. Other species mentioned above are also planted in small quantities or gathered from the forest. The highlanders often gather wild yams and the fruit of an oil *Pandanus* (*P. conoideus*). The oily red fruit is cooked in a stone oven; the pulp and seeds may be sucked off the fruit, or be removed from the stalk, mashed with water and strained through a dry *Pandanus* leaf to produce a thick red liquid. Some green vegetables or bananas may be mixed with it before eating. The addition of salt brings out the flavour.

The highland agriculturalists

In New Guinea nearly 40 per cent of the total population live at altitudes ranging from 1400 to 2700 m in valleys of the central cordillera. The densest populations, locally up to 200 persons per km^2, are to be found in the intermontane valleys and basins, the Wissel Lakes and Baliem areas of Irian Jaya and the Southern, Western, Enga, Chimbu and Eastern Highlands Districts of Papua New Guinea (Brookfield 1961).

Although in some areas of the Eastern Highlands (e.g. Kukukuku, Blackwood 1939; Patep, Conroy and Bridgland 1950) and further west, near Telefomin (Barth 1971) and on the Vogelkop peninsula (Oomen and Malcolm 1958) a fairly simple slash and burn, shifting agriculture is practised, in most areas cultivation practices are characterised by much greater elaboration of land preparation, crop rotation and fallowing than in the lowlands.

Usually the perimeter of an area of forest or forest regrowth chosen for a garden will be cleared first and the fence erected before the undergrowth of the plot is slashed and the larger tree branches pollarded. An outer ditch may be dug to help keep pigs out of the garden. The debris is stacked around the tree bases and burned when dry. Then the stumps of trees, ferns and cane grasses are dug out and the ashes spread over the ground. Taller trees are left for firewood. If an area of grassland is chosen for a new garden this may be simply burnt over, but usually the grass is cut, stacked and burned in isolated piles; the roots are pulled and dug out. The area of ground thus prepared and fenced is then divided up into smaller, individually owned plots. These are cleared further and the ground prepared for planting. This usually involves complete tillage of the plot and often some mounding is undertaken; small mounds of heaped dirt on a base of ashes are found throughout the Eastern Highlands; larger, composted mounds are built in the Western, Enga and Southern Highlands Districts (Plate 46). In the Wahgi valley a system of subsoiling is used: the ground is marked out in square or rectangular blocks and the soil from the perimeter ditches (30 cm wide and deep) of these blocks is thrown on to the plots. On steep slopes log and soil fences may be constructed to control erosion and *Cordyline* sp. and *Casuarina* sp. planted to help hold the soil. In swampy areas deep ditches are dug around the perimeter of the garden and often smaller ditches are dug within the garden to maintain a low water table level. Trees and shrubs are planted along the deep ditch walls to keep them strong, and the lower parts may be vertically planked to stop erosion or damage by pigs.

Gardens may be planted with a mixture of crops or with the staple of the region, the sweet potato (*Ipomoea batatas*). In many areas the first gardens planted following clearance of forest or forest regrowth (Plate 47) are mixed tuber, fruit and green vegetable gardens (taro, yams, beans, sweet potato, bananas, sugarcane, *Setaria palmifolia, Rungia klossii, Oenanthe javanica, Zingiber* sp., etc.), but after these have been harvested sweet potatoes are replanted together with

Plate 46
Sweet potato (*Ipomoea batatas*) garden in the Western Highlands. The three men are standing behind a mound of earth containing compost.

Plate 47
Sugarcane, *Setaria palmifolia*, lima bean, *Oenanthe javanica, Rungia klossii* and sweet potato being grown together in a forest garden at Pureni, Tari Subdistrict

some sugarcane, *Setaria palmifolia* and *Cordyline* sp. The ground is fallowed under grass for short periods (5-10 months) and then reworked for further sweet potato gardens. Once yields decline, the area is left for 5-10 years and woody regrowth may be encouraged by retaining self-sown *Dodonaea viscosa* and *Trema* spp., among others, and planting *Casuarina* sp. seedlings. Stabilised grassland areas are generally used only for sweet potato cultivation. In some areas the mixed crop gardens are maintained separately from the sweet potato gardens; they are used for a period of 2-4 years and then left to develop a woody regrowth over a period of 5-10 years, or longer.

More detailed descriptions of gardening practices can be found in Schindler (1952), Howlett (1962), Brookfield and Brown (1963) and Montgomery (1960) for the Eastern Highlands and Chimbu Districts; in Meggitt (1958a), Clarke (1971), Waddell (1972) and Powell *et al.* (1975) for the Western Highlands; Powell (1973) for the Southern Highlands; and Brass (1941a) and Pospisil (1963) for the Baliem valley and Wissel Lakes areas of Irian Jaya. The ecology of highland agricultural systems is considered in detail by Brookfield (1961, 1962, 1964).

The staple crop — sweet potato. Very many varieties of sweet potato are grown; they differ in leaf size and shape, vine length and growth form, maturity time, tuber size, shape, skin and flesh colour, and cooked flavour and texture. The varieties grown change over the years. Old types are lost and potential new varieties arise all the time as seedlings; these are retained in the gardens and watched and, if considered desirable, reproduced by cuttings. Sweet potato is a highly adaptable species and is tolerant of a wide range of soil and climatic conditions. It prefers light well drained soils but with the specialised tillage methods used (drainage and mounding) can be grown in wet soils and up to 2700 m or more altitude, the upper limit of highland agriculture (Plate 46). At times in such areas drought and frosts may severely damage or even destroy the above-ground parts of the crop but following rain, the underground stems and tubers sprout away again providing planting material for new gardens and also a small crop of poor quality tubers.

Runners are planted, often two to four in a bundle, at any time of the year. Maturity times vary with altitude and soil conditions, some varieties being harvested after 4-5 months on good soil (6-7 months on poor soil), others taking 7-9 or even 12-18 months in some high-altitude areas. Some varieties can produce tubers over a considerable period of time, secondary small tubers being formed after the large primary ones have been harvested. In recent years new varieties of sweet potato have been introduced by Europeans and have been traded throughout the highlands. Some of these are grown mainly as pig food; they may have excellent leaf growth (also used for pig food) and produce a heavy crop of tubers, though these are of poor flavour and eaten by people only in times of scarcity. Sweet potato tubers do not store well, so they are usually harvested when required and cooked the same day; they may be baked in stone ovens or roasted in the ashes.

Sweet potato tubers have recorded yields of 8-15 metric tons per ha in various highland areas (Salisbury 1961; Pospisil 1963; Meggitt 1958a; Rappaport 1967). When grown under experimental conditions on peat soils, individual plants (comprising three cuttings) produced between 3 and 18 kg over three harvests (twelve months) and an average yield of 26-38 metric tons per ha could be expected with high quality mixed varieties; on mineral soils the yields were considerably lower.

Supplementary starchy and other vegetable crops. Supplementary crops in the highlands are usually taro, yams, *Xanthosoma sagittifolium,* bananas, edible grasses (*Saccharum edule* and *Setaria palmifolia*) and beans, sugarcane and many green vegetables (*Rungia klossii, Oenanthe javanica, Solanum nigrum, Amaranthus* spp., *Hibiscus manihot, Rorippa* sp., and *Commelina* spp.). Taro, yams, bananas and the edible grasses are grown in mixed gardens in the same way as in the lowlands. Taro may be cultivated in separate gardens and be irrigated (Conroy and Bridgland 1950) or grown in ditches or swampy areas (Plate 45). It takes 12-18 months to mature in the highlands and yields are not as high as in the lower altitude areas. Yams also take a long time to mature; they are important in some areas but are not often planted above 2300 m altitude. Bananas are considered to provide variety of diet and are important at feasts, especially when pigs are killed. Many varieties are grown; some take 12-18 months to mature, others 2-3 years.

Beans are a seasonal crop of considerable importance in some areas; three types are considered traditional, the hyacinth bean (*Dolichos lablab*), the lima bean (*Phaseolus lunatus*) and the winged bean (*Psophocarpus tetragonolobus*). The last is the most important and is grown widely up to 2200 m altitude. Seeds are planted in separate blocks in mixed gardens and the young plants staked to aid growth. The flowers, leaves, young pods and mature seeds are eaten; so also, in some areas, are the tubers. Many varieties are grown; they differ in flower and vine colour, leaf size, bean pod colour, size, shape and texture, seed size, shape and colour, and cooked flavour. Some varieties are best for tuber production and, in these, the apical leaves and flowers are plucked (as soon as the latter form), so that large bundles of tubers can be harvested after 8-9 months' growth. Other varieties are liked for their high yield of sweet-tasting pods and seeds, harvested after 5-6 months' growth.

Yields of winged bean varieties vary greatly. In recent experiments high quality varieties produced up to 150 g of dry seed and 300 g weight of tubers per plant. With mixed varieties, yields of 1.6-2.2 metric tons per ha of seed and 1.8-2.7 metric tons of tubers were recorded. The nutritional value of these beans is high, as shown by the following percentages: seeds 30-7 protein, 15-19 fat and 28-42 carbohydrate; tubers 12-15 protein, 0.5-1.1 fat and 27 carbohydrate; leaves 6-15 protein, 0.7-1.1 fat; and flowers 6 protein and 0.9 fat.

The hyacinth bean, *Dolichos lablab*, is quite widely grown up to 2100 m altitude, in mixed crop gardens. Seeds are planted near tree trunks so that the vines may climb. A number of varieties are grown; they differ in flower and seed colour, pod shape and colour, cooked flavour and yield. The leaves, young pods and seeds are eaten; the young pods are ready for harvest after 4-6 months' growth, the mature seeds after 6-8 months'. No data on yields are available. The percentage protein content of the immature pods is 4.5 while the dry seeds are 20-8; percentage fat varies from 0.1 in the immature pod to 0.8-1.5 in the dry seeds and carbohydrate from 10 to 57-68 respectively (Powell 1974b). Lima or sieva beans, *Phaseolus lunatus*, are not widely grown and are recorded as traditional only from the Jimi valley and Tari areas of the Highlands. The seeds are planted in mixed gardens in rich, ashy soil near standing trees. The leaves, young pods and seeds are eaten after prolonged cooking in an earth oven. Their chemical composition is similar to that of hyacinth beans.

A wide range of green vegetables is cultivated throughout the highlands. *Amaranthus* spp. are as important as in the lowlands, providing a fast-maturing, pleasant-tasting food; seedlings are thinned and often transplanted in mixed crop gardens. Only a few varieties of *Hibiscus manihot* are grown in the highlands compared with the lowlands and the leaves are attacked severely by insects. *Setaria palmifolia* is widely grown and in some high-altitude areas is a major supplementary crop. Under experimental conditions yields varied from 8-18 kg per plant per annum (seven harvests after an initial five month growth period) depending on variety, or 80-180 metric tons per ha. The edible portion of the varieties varies from 10-18 per cent of stem weight if cooked in bamboo or a saucepan, to about 30 per cent if cooked in an earth oven. The yield of edible portion, then, was 8-49 metric tons per ha.

Rungia klossii is a very important 'spinach' in some areas. It is planted by rooted stems or cuttings in mixed crop gardens and kitchen gardens. The varieties grown differ in leaf size, shape and colour, stem colour, growth form and taste. The plant matures in three to four months and the edible young shoots and leaves may be harvested for twelve months or more before being replanted. Yields varied considerably depending on variety; under experimental conditions most varieties grown produced 1-2 kg of edible leaf and shoot per plant per annum, or 20-40 metric tons per ha. Chemical analyses indicated that the protein content is 2.5-5.0 per cent and the species is relatively rich in vitamins and minerals (Table 3.3). *Rungia klossii* is usualy cooked with *Setaria palmifolia,* but can also be eaten raw. Wild forms growing in the forest are used by hunting parties and sought in times of general scarcity.

Other Acanthaceae, *Hemigraphis* sp. and *Lepidagathis* sp. are cultivated to a lesser extent and gathered also from *Casuarina* groves and forests. The young leaves and shoots are eaten cooked with meat or fish; they are not as sweet or buttery as *Rungia.* A number of varieties of *Oenanthe javanica* are grown; they differ, one from another, in leaf size, petiole and stem colour and growth form. They are reproduced vegetatively by planting cuttings or rooted stems and grown mainly in mixed crop and kitchen gardens. Self-sown varieties are abundant in coffee gardens and *Pandanus* groves, and also in swampland and forest areas; they are gathered by hunting parties. A *Rorippa* sp. is also grown quite widely. Seeds are planted and the whole plant harvested after 1-2 months' growth. It is cooked on its own or with meat, and is soft-textured, with a pleasant flavour. As it is frost-resistant it is a valuable, although subsidiary, food in high-altitude areas. *Commelina cyanea, C. diffusa, Pollia* sp. and *Floscopa scandens* are straggling runners used as vegetables and/or condiments throughout the highlands. They may be planted by cuttings but are often self-sown and left untended in mixed crop and house gardens. Wild forms are used by hunting parties. They are usually cooked with pig. *Solanum nigrum* is a semi-domesticated species, self-sown in mixed crop gardens and areas that have been burnt. The young shoots are eaten raw or cooked; if overcooked they become very bitter. Two domesticated forms known differ in leaf size; a wild form with large seeds is used by hunting parties. Some *Ficus* spp. are cultivated also; the young shoots and fruit of *F. wassa* and *F. copiosa* are eaten, the leaves of *F. dammaropsis* usually cooked with pig and eaten, the fruit being used only in times of scarcity. The leaves and occasionally the fruit of many other *Ficus* spp. are harvested from forest, regrowth and grassland areas (Table 3.1). *Gnetum gnemon* is planted for its edible leaf in *Pandanus-Artocarpus* orchards in the Jimi valley (Clarke 1971).

Fruit and nut trees. *Pandanus* spp. are the most important fruit and nut trees in the highland areas. *Pandanus conoideus,* producing an oily fruit, is grown up to 2300 m altitude. Cuttings are usually planted in groves amongst cane grass and trees, or near houses. The plant is hardy and requires little attention while growing. It matures in about four years. Once the fruit is set the tree's canopy is cut back to increase the sunlight to aid ripening. The fruit may be boiled, roasted or cooked in the ashes; the oily juice is considered a delicacy. In the lower altitude areas are grown many varieties, which differ in plant and leaf size, branching pattern and fruit colour. Wild forms are tended in the forests there also.

Species of nut *Pandanus* (*P. brosimos, P. julianettii*) grow generally between 1700 and 2900 m altitude. There are many varieties. Some are grown in gardens but others are individually owned, tended and/or transplanted in forest groves; the plant bases are kept clear of weeds, and platforms are built around the trunks to keep rats away from the ripening fruit. The plants take 7-8 years to mature, the fruits 2-3 months to ripen. The fruit, a large round ball, comprises hundreds of finger-sized nuts with shells of various degrees of hardness. The kernel and the soft, yellowish pulp at the base are eaten. The fruit may be smoked and stored in the house roof for a long period but it is mostly used as a seasonal food.

Many wild forms occur in the forests also, some types having a hard outer husk, others very small kernels. Clarke (1971) mentions *Pandanus foveolatus* as a high-elevation *Pandanus* with a small kernel, gathered from the forests of the Jimi valley, while Bulmer (pers. comm.) records the use of both young shoots and nuts for food for one particular species in the Kaironk. The nut is an extremely important seasonal supplement in high-altitude areas, and in times of scarcity may become the staple.

Artocarpus altilis, breadfruit, is cultivated up to about 1550 m altitude and wild species are transplanted or tended in the forest. A semi-cultivated species recorded in the Schrader Mountains between 1830 and 2140 m altitude is also used (Bulmer 1964). The seeds rather than the flesh of these species are eaten (Clarke 1971; Sorenson and Gajdusek 1969). Other trees, occasionally cultivated or transplanted at lower elevations in the highlands include *Aleurites moluccana, Mangifera indica* and *Castanopsis acuminatissima*. Fruits are harvested from wild species of *Mangifera* as well. Wild resources include fruits of *Garcinia* sp., *Diospyros* sp., *Syzygium* spp., *Amomum polycarpum* (occasionally transplanted), *Rubus* spp. and *Trichosanthes* sp. and nuts and seeds of *Chisocheton* sp., *Elaeocarpus* spp., *Sloanea archboldiana, Finschia chloroxantha, Sterculia* sp. and *Pittosporum pullifolium* among others (Clarke 1971, Rappaport 1968; Bulmer 1964; Bulmer and Bulmer 1964).

Other subsidiary crops and famine foods. The young fruits of *Lagenaria siceraria* are often eaten; when cooked they are very juicy and sweet. Varieties grown (in both lowland and highland areas) differ in fruit size and shape. Seeds are planted in house or mixed crop gardens and some care is needed to produce an undamaged mature gourd, valued as a container. *Bambusa* spp. are grown widely to provide raw materials for many different things, and some have edible shoots, which are eaten either raw or cooked. They are particularly important in times of scarcity. *Pueraria lobata*, a semi-woody vine with a large edible tuber, is still cultivated in some parts of the highlands today but was probably of greater importance formerly; wild forms are gathered from grassland and forest areas in times of scarcity. The cultivated form produces fibrous tubers, which may be harvested after 12-18 months' growth but are usually left for 3-5 years to mature. They may attain a weight of 20-32 kg

(Watson 1964, 1968; Barrau 1965; Bowers 1964; Strathern 1969; Powell 1974b).

Other wild foods recorded from the highlands include the heart 'cabbage' of a palm, *Gronophyllum chaunostachys,* leaves of *Acalypha* sp. and *Erythrina* sp. (occasionally seedlings are transplanted) and of *Beilschmiedia* sp., *Palmeria* sp., *Astronia* sp., *Boerlagiodendron* sp., *Papuacedrus* sp., *Saurauia* sp., *Rhodomyrtus* sp., *Chloranthus* sp., *Cyrtandra* sp., *Microcos* sp., *Leucosyke* sp., *Pipturus* sp., and *Graptophyllum pictum* among the trees and shrubs; species of *Lactuca, Desmodium, Impatiens, Rubus, Elatostema, Pilea, Laportea, Cypholophus* and *Coleus* among the herbs; leaves and stems of *Begonia* sp. and *Pouzolzia* sp. and stems of *Coix gigantea,* seeds of *Bidens* sp., *Scleria* sp. and *Coix lachryma-jobi.* The young fronds of the ferns *Asplenium* spp., *Blechnum* sp., *Pteris moluccana, Gleichenia* spp., *Thelypteris* sp., *Diplazium* spp., *Dryopteris* sp. and *Polypodium* spp. are also cooked and eaten.

Condiments and flavouring. In most areas of New Guinea the green vegetables are used as 'condiments' to the starchy staple. As well as these the spicy roots of the Zingiberaceae, including *Alpinia* spp., *Amomum* sp., *Curcuma domestica, Zingiber officinale* and other *Zingiber* spp. are used. The use of the young leaves of *Polyscias* sp. and of *Euodia* spp. are recorded for the highlands; the chili, *Capsicum frutescens,* is cultivated on the coast of Irian Jaya and in some parts of Papua New Guinea, and *Kaempferia galanga* and *Ocimum basilicum* have been introduced into some areas (Barrau 1958). The bark of *Cinnamomum* sp. is used widely as a flavouring, and the recently introduced *Allium* spp. are also widely used.

Plant-ash is used as a salt substitute in many areas where salt water is not available. Plants so used include the ferns *Asplenium nidus* and *Polypodium* sp., the ubiquitous grass *Imperata cylindrica,* the swamp herb *Eriocaulon australe* and the balsams *Impatiens platypetala, I. linearifolia, I. mooreana, I. nivea* and *I.* sp.; their use has been recorded from the Jimi valley, Okapa, Menyamya and Telefomin areas of the Highlands of Papua New Guinea and from the foothills of the Finisterre Ranges (Clarke 1971; Dickie and Malcolm 1940; Freund *et al.* 1965). Bamboo sprouts, the whole plant of some Zingiberaceae, stems of *Acalypha insulana* and leaves of *Elatostema macrophyllum, Saurauia* spp. and *Palmeria* sp. are also burnt for salt in some areas (Blackwood 1940).

Coix gigantea is cultivated widely as a source of salt: the ash is used without purification in the Telefomin area but the soluble salt is extracted and crystallised in the Kukukuku area (Freund *et al.* 1965). It is traded and used as currency in the Fore area of the Eastern Highlands (Sorenson and Gajdusek 1969). To produce salt the grass is cut and dried thoroughly and then burnt. The ash is then placed in open-ended gourds with a *Triumfetta nigricans* burr as a lower filter, and water is poured slowly through it, the filtrate running into a bamboo trough and then into bamboo containers. A stove made of stones and earth is used for evaporation; grooves in the upper surface of the stove are lined with fern fronds, flower petals and banana leaves and the solution is poured from the bamboo pipes through the burr filter again into the banana leaves. A constant heat is maintained for 3-5 days to evaporate the water from the salt, further solution being added as the level is reduced. Finally a block of salt is formed and this is carefully wrapped while still warm.

In some areas the gourd funnels are replaced by sewn bark funnels, fern leaflets are used as filters and open-sided lengths of bamboo as evaporating pans. The bamboo containers of dry salt are covered with *Pandanus* sp. leaves and hung up in the roof of the house to smoke. Leaves of the orchids *Spathoglottis* sp. and *Dendrobium* sp. are often used to wrap salt.

The type of salt prepares is usually high in potassium (38 per cent for *Coix gigantea*, 39 per cent for *Imperata cylindrica*), and low in sodium (0.4 and 1.5 per cent respectively). Ash-salt of *Eriocaulon australe*, on the other hand, contains nearly 11 per cent potassium and 20 per cent sodium (Freund *et al.* 1965; Dickie and Malcolm 1940).

A number of plants are used also for obtaining salt from salt springs distributed throughout the country. The process involves steeping dry grasses such as *Imperata cylindrica* or *Agropyron* sp., softwoods or banana stems in brine pools for several months and then drying and burning them. The ash-salt may be used directly but often the salt is purified further by leaching and evaporating as described above (Meggitt 1958a, b; Maahs 1950; Heider 1970). Williams (1930) records for the Orokaiva area coconut husks being burnt with certain leaves for ash-salt, the ash being dissolved in water in a coconut and the salty water used. In the Enga area ash-salt manufactured from a salt spring contained about 2 per cent potassium and 27 per cent sodium (Meggitt 1958a).

Other Useful Plants

Narcotics, stimulants and intoxicants

Tobacco, *Nicotiana tabacum*, is widely grown throughout New Guinea. In some areas it is considered to be truly traditional while in others it is regarded as a relatively recent introduction of pre-European or post-European contact (Williams 1930; Miklouho-Maclay 1885; Clarke 1971; Rappaport 1967). Varieties grown differ in leaf size and shape and pungency. Seeds may be broadcast in garden areas, sown in separate patches of mixed crop gardens or under the eaves of the house. On Frederik-Hendrik Island seeds are sown into banana leaves filled with soil and the seedlings carefully nurtured in shaded nurseries (Serpenti 1965). The leaves are harvested after a few months growth, the midrib pulled off and the leaf surface dried near the fire. While still soft the leaves are rolled up, twisted and stored in palm leaves above the fireplace. The tobacco is rolled in bamboo, banana, taro or *Pandanus* leaves and smoked as a cigarette or used in a bamboo pipe. Leaves of *Macaranga* sp., *Acalypha insulana, Donax canniformis, Kleinhovia hospita, Rubus moluccanus* or *Ficus* spp. are also used as cigarette 'papers' (Pospisil 1963; Floyd 1954; Hogbin 1938-9; Whiting and

Reed 1938; Rappaport 1939; Powell *et al.* 1975). In some areas tobacco is an important trade item (Wedgwood 1934; Reynders 1962).

Chewing of betel nut (*Areca catechu*) as a narcotic is widespread in coastal regions and at lower elevations of the highlands. In some areas both children and adults chew betel, in others only adults are permitted to do so. The *Areca* palm is grown in village groves or as isolated palms near houses. The nuts of wild forms are gathered from the lowland and mountain forests (up to 1000 m altitude), and seedlings are often transplanted into the villages (Hogbin 1938-9; Williams 1930; Blackwood 1940). The nut (actually the hard endosperm of ripe or unripe seeds) may be chewed alone, but usually a quid of the nut, leaves of the betel-pepper (*Piper betle*) and lime are chewed. The bark and inflorescence of the betel-pepper and of other *Piper* spp. may be included with the quid and also, sometimes, tobacco. Lime-substitutes include the bark-ash of *Archidendron* sp., or an unidentified Meliacea (Blackwood 1940). Floyd (1954) records the chewing of the nut of another palm, *Archontophoenix* sp. and of the leaves of *Pueraria phaseoloides* as intoxicants in New Britain, Blackwood (1940) the seeds of *Lactuca indica* among the Kukukuku.

Piper methysticum, kava, is an important narcotic used by the Marind-Anim and Frederik-Hendrik islanders of Irian Jaya and the Kiwai, Keraki and Tugeri people of the Trans-Fly and Gulf areas of south-western Papua New Guinea (Haddon 1916; Serpenti 1965); it has been recorded as traditional also for the Rai coast people by Miklouho-Maclay (1885) and in the Admiralty Islands (Chowning, pers. comm.). The plant, a small shrub, is grown from cuttings in house gardens or small groves. The chewed, fresh root is spat into coconut shell bowls, mixed with water and strained through a coconut palm leaf-sheath, the filtrate being drunk. In some areas the leaves and stem are chewed as well as the root and in others dried roots are chewed to produce an extremely bitter drink. Kava acts as a sedative, soporific and hypnotic and is said to bring about pleasant dreams (Purseglove 1968). *Kaempferia galanga* is used to induce pleasant dreams and as a hallucinogen ('dream man') in the Morobe and Fore areas of Papua New Guinea respectively (Henty 1960).

The sap of various palms, including *Arenga saccharifera*, *Nypa fruticans* and *Cocos nucifera*, provides an important food and intoxicating drink in the Waropen and Vogelkop peninsula areas of Irian Jaya (Barrau 1958; Held 1957; Oomen and Malcolm 1958). The starch stored in the stem is converted into sugar when the palm begins to flower and this is tapped by bruising and cutting the peduncle or spathe of the inflorescence and collecting the sap in bamboo tubes or gourds. The unfermented sweet sap may be drunk but fermentation occurs rapidly, owing to the action of naturally occurring yeasts, to produce a toddy (Purseglove 1972). Held (1957) notes that fermentation of *Nypa* palm sap is started by adding a portion of the aerial root of a mangrove in the Waropen area.

Medicinal plants

Almost all cases of sickness, injury and death are attributed to supernatural powers. Even minor accidents, cuts and bruises, toothache and general body pains may be caused by such outside influences. Glick (1972) distinguishes between *ailments,* which are socially insignificant and usually treated in a simple, practical way, and *illnesses,* which are much more serious socially; they require explanation (in terms of sorcery and witchcraft) and demand responses which involve the use of ritual and magic. Medicinal plants are involved in both cases but are considered to play a far less important part in effecting cures in the case of illnesses. Many anthropological studies record ritual procedures involved in treating ailments and illnesses and consider these in relation to sorcery, witchcraft, religion and magic (Wedgwood 1934; Williams 1930, 1936; Hogbin 1938-9; Luzbetak 1958; Glick 1967; Schiefenhoevel 1971). Recently attention has been turned to the plants themselves, to those used both in treating simple ailments and in effecting cures of illnesses (Floyd 1954; Stopp 1963; Straatmans 1971; Panoff 1970a; Womersley 1973) and some analyses of the chemical properties of them have been undertaken (Webb 1955, 1959; Hartley 1973; Holdsworth 1973, 1974a, b; Holdsworth and Heers 1971; Holdsworth and Longley 1972; Holdsworth and N'Drawii 1973; Holdsworth and Tringen 1973). Tables 3.5 to 3.13 list some of the plants utilised as medicines in Papua New Guinea; no information is readily available for Irian Jaya.

Cuts and wounds. Cuts and wounds may be bathed with the cotton of *Hibiscus tiliaceus* (in coastal areas) or treated directly with antiseptic solutions. These are made by simply crushing the stem,

Table 3.5 Plants used in the treatment of cuts and wounds

Plant name	Family	Part used and treatment	Area recorded
† *Alpinia* sp.	Zingiberaceae	Rhizome sap applied	Mt Hagen
* *Calophyllum inophyllum*	Clusiaceae	Leaves heated, applied	Manus
Chenopodium sp.	Chenopodiaceae	Leaves (?), stop bleeding	—
Cordyline fruticosa	Liliaceae	Leaves, stems heated, applied	Manus, Central D.
Dendrocnide excelsa (syn. *Laportea gigas*)	Urticaceae	Bark, mixed with *Erythrina indica* bark, applied	New Britain
Eleusine indica	Poaceae	Stops bleeding	Northern D.
Gynostemma pentaphylla	Cucurbitaceae	Sap applied	New Britain
Hibiscus sp.	Malvaceae	Pod cotton, bathe wounds	New Britain
Hyptis sp.	Lamiaceae	Leaves applied	d'Entrecasteaux
Impatiens sp.	Balsaminaceae	Plant sap applied hot	—
* *Imperata cylindrica*	Poaceae	Pith rubbed in, cover with leaves	Tari
* *Merremia peltata*	Convolvulaceae	Stem crushed, applied	Manus
Monstera sp.	Araceae	Stem sap heated, applied	New Britain
Morinda citrifolia	Rubiaceae	Root bark infusion applied	—
* *Paspalum conjugatum*	Poaceae	Flower bud crushed, applied	Sepik
Paspalum sp.	Poaceae	Leaves chewed, applied	Central D.
Pothos sp.	Araceae	Stem, leaves crushed, applied	New Britain
Rhaphidophora sp.	Araceae	Stem crushed, sap applied	—
Rubus sp.	Rosaceae	Berry sap rubbed in	Baliem
Sida rhombifolia	Malvaceae	Sap applied to stop bleeding	New Britain
Vatica papuana	Dipterocarpaceae	Leaves (?), stop bleeding	—
Wedelia biflora	Asteraceae	Stem heated, crushed, applied	Sepik, Manus
		Leaf sap applied to stop bleeding	Central D.
* *Zingiber* sp.	Zingiberaceae	Rhizome heated, sap applied	d'Entrecasteaux
		Stops bleeding	Northern D.

* Species which are used also in the treatment of sores.
† Species used also in the treatment of burns.

leaves, flowers or roots of specific plants in water (eight species recorded, Table 3.5) or by heating the whole plant or plant part before crushing and applying the solution (four species); leaves or bark may be applied directly also (five species). Excessive bleeding may be stopped by using *Chenopodium* sp., *Vatica papuana* (Womersley 1973) or the sap of *Sida rhombifolia, Wedelia biflora* or *Zingiber* sp. Occasionally plant substances, for example the pith of the kunai grass, are rubbed into cuts; *Imperata cylindrica* is used thus in the Tari area, and fruit of the wild strawberry, *Rubus* sp., in the Baliem valley of Irian Jaya (Table 3.5). Leaves of *Blumea* sp. are rubbed on the broken skin of pigs to aid healing in the Tari area.

Burns. Burns may be treated with masticated bark of taun, *Pometia pinnata,* or *Canarium indicum* (*C. commune*) in coastal areas of New Britain (Floyd 1954), with *Pandanus* sp. fruit juice mixed with lime, or resin of *Elmerrillia papuana* mixed with the fluid of cooked sago grubs in the Bosavi area of the Southern Highlands (Schiefenhoevel 1971). Widespread treatments for burns include use of crushed gingers (*Alpinia* sp. and *Zingiber zerumbet*), the application of shredded taro leaves or of the tuber sap of sweet potato (Table 3.6).

Table 3.6 Plants used in the treatment of burns*

Plant name	Family	Part used and treatment	Area recorded
Canarium indicum (syn. *C. commune*)	Burseraceae	Bark masticated, applied	New Britain
Colocasia esculenta	Araceae	Shredded young leaves applied	New Britain, Mt Hagen, Tari
Elmerrillia papuana	Magnoliaceae	Resin mixed with fluid from cooked sago grubs, applied	Bosavi
Impatiens mooreana	Balsaminaceae	Young leaves, petals applied	Mt Hagen
Ipomoea batatas	Convolvulaceae	Tuber sap applied	—
Pandanus sp.	Pandanaceae	Fruit juice mixed with lime applied	Bosavi
Pometia pinnata	Sapindaceae	Bark masticated, applied	New Britain
Zingiber zerumbet	Zingiberaceae	Rhizome crushed, applied	New Britain

* See Table 3.2 also.

Table 3.7 Plants used in the treatment of sores*

Plant name	Family	Part used and treatment	Area recorded
Ageratum sp.	Asteraceae	—	Manus
Artocarpus sp.	Moraceae	Latex applied	New Britain
Athyrium sp.	Athyriaceae	—	Northern D.
Bidens pilosa	Asteraceae	Seeds used to extract pus	Tari
Biden sp.	Asteraceae	Latex diluted, applied	New Britain
Bridelia minutiflora	Euphorbiaceae	Leaves heated, applied	Central D.
Bryophyllum sp.	Crassulaceae	Leaves heated, applied	—
Carica papaya	Caricaeae	Latex	Northern D., d'Entrecasteaux
Centella asiatica	Apiaceae	Plant powdered, mixed with lime, applied	Sepik, Manus, Central D.
Cerbera manghas	Apocynaceae	Latex applied	New Britain
Cissus sp.	Vitaceae	—	Northern D.
Clerodendron sp.	Verbenaceae	Heated leaves applied	Central D.
Cocos nucifera	Palmae	Leaf chewed with scraped coconut, applied	Central D.
Codiaeum variegatum	Euphorbiaceae	Leaf sap applied	Central D., Manus
Crassocephalum crepidioides	Asteraceae	Leaves, petioles heated, crushed, applied to abscesses	Mt Hagen
Cycas circinalis	Cycadaceae	Ovule crushed, applied	Central D.
Cycas spp.	Cycadaceae	Ovule powdered, lime added, applied	New Britain, Manus, Central D.

Plant name	Family	Part used and treatment	Area recorded
Cyclosorus sp.	Thelypteridaceae	Leaves boiled, applied	New Britain
Diospyros sp.	Ebenaceae	Leaves masticated, applied	Jimi
Dodonaea viscosa	Sapindaceae	Leaf sap applied	Wahgi, Baliem
Emilia sp.	Asteraceae	Leaves heated, applied	Ialibu
Ficus botryocarpa	Moraceae	—	Northern D.
Ficus pachyrrhachis	Moraceae	—	Northern D.
Ficus sp.	Moraceae	Latex of fruit applied	New Britain
Gmelina sp.	Verbenaceae	Bark powdered, applied	New Britain
Hibiscus tiliaceus	Malvaceae	Leaves heated, applied	Central D.
Hypericum papuanum	Hypericaceae	Leaves	Central D.
Ipomoea congesta	Convolvulaceae	Leaves boiled, sap applied	New Britain, Northern D.
Impomoea pes-caprae	Convolvulaceae	Leaves heated, applied	Manus
		Apical shoot sap applied	Central D.
Kalanchoe pinnatum	Crassulaceae	Leaves heated, applied	Manus, Central D.
Lactuca sp.	Asteraceae	Flowers, seeds crushed, sap applied	Tari
Macaranga sp.	Euphorbiaceae	Leaves heated, applied	Manus
Microglossa pyrifolia	Asteraceae	—	Northern D.
Morinda sp.	Rubiaceae	Heated leaves rubbed on	Central D., d'Entrecasteaux
Nicotiana tabacum	Solanaceae	—	Northern D.
Oenanthe javanica	Apiaceae	Leaves crushed, applied	Southern Highlands D.
Omalanthus sp.	Euphorbiaceae	Sprouts crushed, sap applied	Mt Hagen
Pandanus sp.	Pandanaceae	—	Trobriands
Pangium edule	Flacourtiaceae	—	Northern D.
Papuechites aambe	Apocynaceae	Stem sap applied	Manus
Parartocarpus venenosa	Moraceae	Seeds powdered, mixed with lime, applied	New Britain
Pennisetum sp.	Poaceae	—	Northern D.
Plectranthus scutellarioides	Lamiaceae	Leaves rubbed on	Tari
Premna integrifolia	Verbenaceae	Leaves masticated, applied	Manus
Pterocarpus indicus	Leguminosae	Boiled leaf solution applied	Central D., d'Entrecasteaux
Sacciolepis indica	Poaceae	Shoot sap applied	Tari
Scaevola taccada	Goodeniaceae	Leaves applied	—
Syzygium sp.	Myrtaceae	Bark masticated, applied	New Britain
Ternstroemia cherryi	Theaceae	Seeds powdered, applied	Manus
Tridax procumbens	Asteraceae	Leaf sap applied	Central D.
Viola sp.	Violaceae	Leaves crushed, applied	Mt Hagen
Zingiber sp.	Zingiberaceae	Rhizome masticated, applied	New Britain

* See Table 3.2 also.

Skin diseases and sores. Skin diseases such as dermatitis and grille, sores and tropical ulcers are widespread in New Guinea and many different plants are used to treat them. Fifty-seven species are recorded for the treatment of sores and ulcers (Tables 3.5 and 3.7). In many cases the treatment given is similar to that used for cuts and wounds, namely the application of the sap of raw or cooked leaves, stems and flower buds (sixteen species), or cold or heated leaves may be applied directly (fourteen species); in other cases leaves, bark or rhizomes are masticated and the mixture of sap and saliva applied (five species). Latex of breadfruit trees, fig trees, papaya, and *Cerbera manghas* and *Bidens* sp. aids healing. Sores and ulcers may be dried up by applying (either directly or mixed with lime) preparations of powdered seeds and ovules of *Parartocarpus venenosa*, *Ternstroemia cherryi* and *Cycas* spp. respectively; powdered bark of *Gmelina* sp. is also effective, and for babies' sores the whole plant of *Centella asiatica*, ground up with lime, is used (Floyd 1954; Panoff 1972; Webb 1960).

Dermatitis and grille are treated with the bark of *Croton* sp. or *Dolichandrone spathacea*, with crushed young leaves of *Cassia alata* or cooked leaves of *Commelina* sp. In the Orokaiva area the latex of a *Ficus* sp., added to scrapings of yam tubers and wallaby blood, is used (Williams 1930); in the Bosavi area of the Southern Highlands the dark latex of *Semecarpus* sp. (Schiefenhoevel 1971). Ringworm is treated in New Britain (and probably elsewhere) with a mixture of lime and leaves of either *Ocimum sanctum* or *Cucumis* sp. (Panoff 1972); scabies is treated with masticated leaves of *Rubus rosifolius* in the Jimi valley (Clarke 1971), with *Saccharum spontaneum* or the endosperm of the coconut in the Northern District, and with *Psidium guajava* on Manus (Holdsworth 1974a); leprosy is treated with the petiole sap of a tall fern, *Marattia* sp. in the Hagen area (Stopp 1963), with *Lygodium* sp. in the d'Entrecasteaux Islands or with *Balanophora* sp. elsewhere (Womersley 1973). Other skin rashes may be treated with boiled leaf solutions of *Hibiscus manihot* or *Calophyllum inophyllum* or *Syzygium* sp. in the Central District and d'Entrecasteaux Islands respectively (Holdsworth 1974a, b).

Itches are eased with the leaf sap of *Begonia angustae* or with *Maesa* sp., stings with the cooked leaves of *Coleus* sp., the sap of crushed *Piper* sp. stems or the latex of *Excoecaria* sp. Few treatments of snake bites are recorded: the sap of crushed ginger (*Zingiber* sp.) or the latex of *Plumeria acutifolia*, rubbed into the bite, and the bark of *Mangifera minor* masticated and swallowed, are all considered to be effective. Holdsworth and Farnworth (1974) record also the application of crushed leaves of *Derris* sp. or *Timonius timon* mixed with coconut water, and the ritual use of leaves of *Barringtonia* sp. from the Central District. Pig bites may be treated with crushed leaves of *Psychotria* sp. and millipede bites rubbed with the soft bark of the trunk of bananas (*Musa* spp.).

Dressings and bandages. Many plants are available for use as dressings and bandages but few are recorded in the literature. Noted from New Britain are *Cordyline* sp. and *Inocarpus fagiferus* leaves as dressings, *Costus speciosus* and *Tapeinocheilos dahlii* stems as splints. In the Bosavi area fronds of *Angiopteris evecta* are used as splints, bound with the fibre of *Hibiscus tiliaceus*. In the Mt Hagen area the leaf base of *Crinum macrantherum* is used as a dressing for stab and arrow wounds.

General body pains. The plants used to relieve general body pains, aching muscles and joints and headaches are listed in Table 3.8. Treatments often involve rubbing the body or aching part with liniments (*Alphitonia* spp., *Dodonaea viscosa, Endospermum formicarum*) or prepared mixtures of lime, clay or soil with plant products (*Centella asiatica, Wendlandia paniculata, Coleus* sp., *Cordyline* sp.). In other cases cold or heated leaves, bark or mixed preparations are applied to the aching part, or counter-irritants such as *Fleurya* sp. and *Laportea* spp. are used; various *Alpinia* and *Zingiber* species masticated with ginger are also considered to be effective.

For headaches, further treatments include eating the fruit of *Melochia tomentosa*, or taking extracts of *Phyllanthus amarus, Pterocarpus indicus, Ficus septica, Premna integrifolia* or *Peripterygium moluccanum* (*Cardiopteris moluccana*) leaves, stems or flowers, bathing the head with a solution of *Ageratum conyzoides*

Table 3.8 Plants used to relieve headaches, general body pains and swellings

Plant name	Family	Complaint	Part used and treatment	Area recorded
Ageratum conyzoides	Asteraceae	Headache	Bathe with leaf extract	Manus
Alphitonia incana	Rhamnaceae	Sore hands	Leaves rubbed in	Baliem
		Swellings	Bark applied	Mt Hagen
Alphitonia moluccana	Rhamnaceae		Linament	—
Alpinia sp.	Zingiberaceae	Headache, body pain	—	Northern D., Kukukuku
Alstonia brassii	Apocynaceae	Headaches	—	Northern D.
Alstonia scholaris	Apocynaceae	Abdominal, chest pains	—	Northern D.
Artocarpus altilis	Moraceae	Swollen groins	Leaves crushed with *Carica papaya*, leaves and lime, applied	d'Entrecasteaux
Begonia sp.	Begoniaceae	Abdominal pain	Leaves heated, rubbed on skin	Watut, EHD
Boehmeria sp.	Urticaceae	Headaches	Leaves rubbed on forehead	d'Entrecasteaux
Callicarpa sp.	Verbenaceae	Body pain	—	Northern D.
Cardiopteris moluccana	Cardiopteridaceae	Headache	Fronds crushed, mixed with water, lime juice, drunk	Sepik
Carica papaya	Caricaceae	Headaches	Latex	Northern D., d'Entrecasteaux
Centella asiatica	Apiaceae	Muscle, joint pains	Plant mixed with lime, *Wendlandia* sp., applied	Trobriands
Cinnamomum sp.	Lauraceae	Chest pain	Leaves masticated, applied	—
Coleus spp.	Lamiaceae	General body pain	Leaves mixed with soil, rubbed in	Kukukuku, Watut
		Muscular pain	Leaves masticated, applied	—
Cordyline sp.	Liliaceae	General	Leaves mixed with clay, rubbed in	Kukukuku
Cymbopogon sp.	Poaceae	Headache	—	—
Dioscorea sp.	Dioscoreaceae	Body pains	—	Northern D.
Dodonaea viscosa	Sapindaceae	Aching shoulders	Leaves rubbed in	Baliem
Elatostema sp.	Urticaceae	Swellings	Leaves applied	—
Endospermum formicarum	Euphorbiaceae	General	Leaves, bark heated, rubbed in	New Britain
Epipremnum sp.	Araceae	Chest pains	—	Northern D.
Euodia sp.	Rutaceae	Body pain	—	Northern D.
Eurycles amboinensis	Amaryllidaceae	General	Leaves masticated with ginger, applied	Kukukuku
Ficus pungens	Moraceae	Body pains	—	Northern D.
Ficus septica	Moraceae	Swellings, bruises	Leaves mixed with *Macaranga* sp., applied	Trobriands
		Headache	Leaf, stem sap, diluted, drunk	d'Entrecasteaux
Ficus sp.	Moraceae	Headache	Pressure applied and released alternately with leaves	Kukukuku
Fleurya sp.	Urticaceae	General	Leaves as counter-irritant	Widespread
Impatiens sp.	Balsaminaceae	Head, chest pain	Aqueous plant extract drunk	Watut, EHD
Kalanchoe pinnatum	Crassulaceae	Headache	Leaves heated, applied to forehead	Manus
Laportea spp.	Urticaceae	General pain, headaches, muscular fatigue	Leaves as counter-irritant	Widespread
Leucosyke candidissima	Urticaceae	Abdominal pain	Leaves heated, pressed on skin	Watut, EHD
Macaranga sp.	Euphorbiaceae	Headache, general	Leaves heated, applied to pain	Manus
		Swellings, bruises	Leaves used with *Ficus septica* in blood-letting	Trobriands
Melochia tomentosa	Sterculiaceae	Headache	Fruit eaten, leaves applied to forehead	—
Metroxylon spp.	Palmae	Headache	Stem sap rubbed on forehead	d'Entrecasteaux
Morinda citrifolia	Rubiaceae	Headache, general	Leaves heated, applied to pain	Manus
Musa spp.	Musaceae	Body pains	—	Manus
Mussaenda ferruginea	Rubiaceae	Headaches	—	Northern D.
Oenanthe javanica	Apiaceae	Headache	—	Tapini
Philodendron sp.	Araceae	Swellings	Young leaves mixed with lime, applied	d'Entrecasteaux
Phyllanthus amarus	Euphorbiaceae	Swelling, headache	Whole plant boiled, extract drunk	Sepik
Pipturus repandus	Urticaceae	Swelling	—	—

Polystichum sp.	Aspidiaceae	Groin swelling	Frond heated, crushed, applied	Mt Hagen
Premna integrifolia	Verbenaceae	Headache	Boiled leaf solution	d'Entrecasteaux
Pterocarpus indicus	Leguminosae	Headache	Flowers boiled, solution drunk	Central D.
Rubus glomeratus	Rosaceae	Abdominal pain	Leaves heated, pressed on skin	Watut, EHD
Wedelia biflora	Asteraceae	Headache	Leaves dried, crushed with lime, applied to forehead	Northern D., Sepik, Manus
Wendlandia paniculata	Rubiaceae	Muscle, joint pains	Leaves mixed with lime, *Centella asiatica*, applied	Trobriands
Zingiber zerumbet	Zingiberaceae	General	Rhizome masticated, rubbed in	Widespread

EHD = Eastern Highlands District

Table 3.9 Plants used to relieve toothache and other mouth infections

Plant name	Family	Part used	Area recorded
Alpinia sp.	Zingiberaceae	Rhizome	Mt Hagen
Areca catechu	Palmae	Nut	New Britain
Blumea sp.	Asteraceae	Plant	Mt Hagen, Baliem
Borreria laevis	Rubiaceae	Leaves	SHD
Desmodium sequax	Leguminosae	Root	SHD
Drimys piperita	Winteraceae	Fruit exocarp	Tari
Emilia prenanthoidea	Asteraceae	Leaves	Ialibu
Euphorbia plumerioides	Euphorbiaceae	Latex	Mt Hagen
Impatiens sp.	Balsaminaceae	Plant	Watut, EHD
Ocimum basilicum	Lamiaceae	Leaves	Central D.
Osbeckia sinensis	Melastomataceae	Plant	Mt Hagen
Pennisetum sp.	Poaceae	—	Northern D.
Pipturus argenteus	Urticaceae	Sap	Mt Hagen
Pipturus repandus	Urticaceae	Leaves	Sepik
Pittosporum ferrugineum	Pittosporaceae	Root bark	Central D.
Psychotria sp.	Rubiaceae	—	Northern D.
Pteridium aquilinum	Dennstaedtiaceae	Petiole sap	Mt Hagen
Sida cordifolia	Malvaceae	—	Central D.
Sida rhombifolia	Malvaceae	—	Northern D.
Timonius sp.	Rubiaceae	—	Northern D.
Viola klossii	Violaceae	Sap	Mt Hagen
Zingiber zerumbet	Zingiberaceae	Rhizome sap	SHD

EHD = Eastern Highlands District
SHD = Southern Highlands District

or applying presure to and releasing it from the head alternately using a *Ficus* sp. leaf.

Swellings are treated with the bark of *Alphitonia incana*, leaves of *Elatostema* sp., *Pipturus repandus*, and of *Philodendron* sp. mixed with lime, or by blood-letting involving the use of *Macaranga* sp. and *Ficus septica* to staunch the flow of blood. Groin swellings are treated with hot, crushed fronds of the fern *Polystichum* sp., or with a mixture of breadfruit, papaya leaves and lime (Table 3.8).

Toothache. Plants used to relieve toothache are listed in Table 3.9. In most cases the treatment involves the mastication of the plant species with ginger and traditional ash-salt (*Blumea* sp., *Osbeckia sinensis*) or crushing the mixture over the painful tooth (*Borreria laevis, Desmodium sequax, Emilia prenanthoidea*). In other cases the sap of a specific plant is applied to the tooth; *Pipturus argenteus, Viola klossii, Pteridium aquilinum, Euphorbia plumerioides* are used in this way. The fruit exocarp of *Drimys piperita* inserted into the tooth cavity deadens pain for a considerable period of time. Betel nut (*Areca catechu*), may be chewed on its own or with leaves of *Pipturus repandus* to cure mouth ulcers.

Ear and eye infections. To relieve earache and to treat ear and eye infections plant solutions are usually applied to the affected part, for example, in the Bosavi area of the Southern Highlands clear infusions of *Cayratia* sp. or *Calamus* cf *hollrungii* are applied to ear infections, in the Mt Hagen area the sap of *Hyptis* sp. and of *Rhipogonum* sp. is used, and in New Britain the sap of *Smilax* sp. and *Scaevola* spp. are important throughout the country: the sap of *S. oppositifolia* is applied to ear infections in the Mt Hagen area, the steam produced by boiling crushed leaves of *S. taccada* is used for both eye and ear troubles on Manus, and a salt water extract of its crushed leaves used to bathe sore eyes in the Trobriands. The sap of *Dolichandrone spathacea* mixed with lime is used to treat conjunctivitis in New Britain, while in the d'Entrecasteaux Islands latex of papaya leaf petioles or the sap of *Ipomoea pes-caprae* may be applied; in the Tari area of the Southern Highlands, *Commelina diffusa* is used to clear the eyes of dirt.

Fevers. Table 3.10 lists twenty-five plants used in treating fevers, including malaria, in various parts of the country. The methods employed are essentially the same as those described for headaches and general body pains. Plant products are rubbed on to the skin and plant extracts are either taken internally or used to bathe the

Table 3.10 Plants used in the treatment of fevers, including malaria

Plant name	Family	Part used and treatment	Area recorded
Aristolochia sp.	Aristolochiaceae	—	Northern D.
Bidens sp.	Asteraceae	Latex diluted, drunk	New Britain
Cinnamomum spp.	Lauraceae	Bark applied	—
Codiaeum variegatum	Euphorbiaceae	Body bathed with leaf extract	Manus
Coleus scutellarioides	Lamiaceae	Leaves, flowers rubbed on skin	Kukukuku
Cordyline sp.	Liliaceae	Leaves mixed with clay, rubbed in	Kukukuku
Cryptocarya sp.	Lauraceae	—	—
Cyclosorus sp.	Thelypteridaceae	Plant crushed, extract drunk	Mt Hagen
Euodia anisodora	Rutaceae	—	—
Euodia sp.	Rutaceae	—	Northern D.
Ficus septica	Moraceae	Leaf and stem sap diluted, taken as preventive dose	d'Entrecasteaux
Hypericum japonicum	Hypericaceae	Plant eaten with ginger and ash-salt	Mt Hagen
Morinda citrifolia	Rubiaceae	Leaf sap diluted, drunk	d'Entrecasteaux
Musa spp.	Musaceae	Banana crushed, coconut milk added, eaten	Ramu valley
Mussaenda ferruginea	Rubiaceae	—	Northern D.
Ocimum basilicum	Lamiaceae	Plant boiled, body heated over steam	Central D.
Phyllanthus urinaria	Euphorbiaceae	—	—
Pongamia pinnata	Leguminosae	Crushed leaf extract drunk, body bathed in solution	Sepik
Premna integrifolia	Verbenaceae	Stem, leaves boiled, steam inhaled, body bathed, leaves rubbed into skin	Sepik
		Cold leaf sap diluted, drunk	d'Entrecasteaux
Rhipogonum sp.	Liliaceae	—	—
Securinega virosa	Euphorbiaceae	Boiled leaf solution drunk	Central D.
Stephania sp.	Menispermaceae	Shredded root boiled, steam inhaled and body heated over steam	Central D.
Timonius timon	Rubiaceae	Leaves eaten raw or boiled, patient bathed with solution	Central D.
Tournefortia sarmentosa	Boraginaceae	Leaf chewed, swallowed	Central D.
Villebrunea rubescens (syn. *Oreocnide rubescens*)	Urticaceae	—	Northern D.

body, or both. As well, the body may be heated with steam to induce sweating.

Coughs, colds and sore throats. Forty species are recorded as being used for the treatment of coughs, colds and sore throats (Table 3.11). Treatment in most cases involves drinking plant extracts, such as the diluted sap of *Calamus* sp., diluted latex of members of the Moraceae and Apocynaceae, aqueous extracts of crushed leaves of *Claoxylon* sp., *Premna integrifolia, Erythrina variegata, Euodia hortensis, Scaevola taccada* and *Hibiscus tiliaceus*. The mucilaginous leaves and shoots of Aibika, *Hibiscus manihot*, are widely

Table 3.11 Plants used in the treatment of coughs and colds, sore throats

Plant name	Family	Complaint	Part used and treatment	Area recorded
Alstonia scholaris	Apocynaceae	Cough	Stem sap diluted, drunk	d'Entrecasteaux
Alstonia spectabilis	Apocynaceae	Cough	Leaves crushed, boiled, drunk	Central D.
			Stem sap diluted, drunk	d'Entrecasteaux
Alstonia sp.	Apocynaceae	Cold	Latex diluted, drunk	New Britain
Averrhoa carambola	Oxalidaceae	Sore throat	—	Sepik
Bixa orellana	Bixaceae	Colds in children	—	—
Calamus sp.	Palmae	Colds	Sap diluted, drunk	New Britain
Claoxylon sp.	Euphorbiaceae	Cough	Leaves crushed, diluted, drunk	d'Entrecasteaux
Clematis clemensiae	Ranunculaceae	Cold	—	Northern D.
Clematis glycinoides	Ranunculaceae	Colds	Leaves crushed, inhaled	Widespread
Clematis sp.	Ranunculaceae	Colds	Cooked leaves eaten, steam inhaled	New Britain
Coleus scutellarioides	Lamiaceae	Colds	—	Northern D., New Britain
Coleus sp.	Lamiaceae	Cold	Aqueous leaf solution inhaled	Watut, EHD
Colocasia esculenta	Araceae	Sore throat	Leaves eaten	New Britain
Cyclosorus sp.	Thelypteridaceae	Cough	—	Northern D.
Dendrobium sp.	Orchidaceae	Cough	Leaves	Central D.
Dioscorea sp.	Dioscoreaceae	Cough	—	Northern D.
Elatostema sp.	Urticaceae	Cough	Masticated with traditional salt	Watut, EHD
Erythrina variegata	Leguminosae	Cough	Leaves crushed in water, drunk	Sepik
Euodia altata	Rutaceae	Cough	—	Northern D.
Euodia hortensis	Rutaceae	Cold	Leaves crushed, diluted, drunk	Sepik
Euodia sp.	Rutaceae	Whooping cough	—	Northern D.
Ficus pungens	Moraceae	Cough	Root crushed, sap diluted, drunk	Sepik
Ficus septica	Moraceae	Cough	Apical leaves eaten with ginger	Central D.
Ficus sp.	Moraceae	Cough	—	Northern D.
Hibiscus manihot	Malvaceae	Cold	Leaves, shoots cooked, soup drunk	Ramu
		Sore throat	—	New Britain
Hibiscus tiliaceus	Malvaceae	Sore throat	Leaves crushed, diluted, drunk	Sepik
		Severe cough	Boiled leaf extract drunk	Central D.
			Leaves crushed, diluted, drunk	Manus
Kleinhovia hospita	Sterculiaceae	Cough	Bark crushed, diluted, drunk	Manus
Octamyrtus sp.	Myrtaceae	Cough	—	Northern D.
Parsonsia sp.	Apocynaceae	Cold	Sap drunk	—
Peperomia sp.	Piperaceae	Cold	—	Northern D.
Polypodium sp.	Polypodiaceae	Catarrh	Plant burned, smoke inhaled	Mt Hagen
Premna integrifolia	Verbenaceae	Cough	Boiled leaf solution drunk	d'Entrecasteaux
Saccharum officinarum	Poaceae	Sore throat	Sugarcane eaten	New Britain
Scaevola taccada	Goodeniaceae	Cough	Leaves crushed in salt water, drunk	Manus
Symbegonia sp.	Begoniaceae	Cough	Leaf boiled with food, eaten	SHD
Syzygium spp.	Myrtaceae	Sore throat	Bark masticated	New Britain
		Cough	Leaves eaten	Kukukuku
Wedelia biflora	Asteraceae	Cough, sore throat	Stem crushed, diluted, drunk	Sepik, Northern D.
Zingiber zerumbet	Zingiberaceae	Cough	Rhizomes masticated with ash-salt	Widespread

EHD = Eastern Highlands District
SHD = Southern Highlands District

used for curing both colds and sore throats; cooked leaves are eaten and the soup drunk. The pulpy cooked leaves and petioles of taro, *Colocasia esculenta,* are also used in this way, and sugarcane, *Saccharum officinarum,* is eaten often to 'cool' the throat. Leaves of *Syzygium* sp. are eaten or the bark of it or ginger, *Zingiber zerumbet,* masticated (the latter with traditional ash-salt) to cure sore throats. For colds and catarrh the leaves of *Coleus* sp. and of *Clematis* spp. may be crushed and inhaled or the latter cooked and the steam inhaled; the inhaled smoke of the fern *Polypodium* sp. is also considered effective in the Mt Hagen region of the Highlands.

Respiratory complaints and tuberculosis. Fifteen species are recorded as useful in treating respiratory complaints. *Coleus* sp., *Laportea* sp., *Cordyline* sp. and *Euphorbia hirta* are used for breathing difficulties, *Piper* sp. and *Albizia falcataria* to clear congestion (the latter causing vomiting) and *Cinnamomum culilawan* for respiratory complaints in general. Extracts of leaves of *Clerodendron* sp. mixed with *Morinda citrifolia* or *Blumea* sp. are taken in the Trobriands for pneumonia and tuberculosis respectively, while extracts of the young leaf tips of *Pterocarpus indicus* and of the bark of *Kleinhovia hospita* or *Hibiscus tiliaceus* are used in the Central District and on Manus respectively to treat tuberculosis (Holdsworth and Farnworth 1974; Holdsworth and Heers 1971). The young flower buds of *Musa* sp. are eaten in the Mt Hagen area to cure phthisis (Stopp 1963). Bronchitis is treated in the Central District with a concoction of the roots of *Erythrina variegata* (Holdsworth and Farnworth 1974).

Stomachaches, dysentery and diarrhoea. Table 3.12 lists fifty-four genera of plants, representing thirty-four families, used in the treatment of stomach complaints. Treatments involve drinking extracts and chewing or eating leaves, stems, roots or fruits of specific plants. The use of domesticated plants is recorded widely; for example, a soup prepared from banana leaves and shoots is drunk, or bananas eaten with *Hibiscus manihot* and *Phyllanthus* sp., or sugarcane or the cooked leaves and petioles of taro eaten. The fermented sap of another Aracea, *Amorphophallus campanulatus,* may be drunk and the young leaves of tobacco, *Nicotiana tabacum,* are chewed and swallowed. Three members of the Palmae are

Table 3.12 Plants used in the treatment of dysentery, diarrhoea and stomachaches

Plant name	Family	Part used and treatment	Area recorded
Adenanthera pavonina	Leguminosae	—	—
Ageratum conyzoides	Asteraceae	—	Northern D.
Alstonia scholaris	Apocynaceae	Bark shredded, diluted extract drunk	New Britain, Sepik, Manus
Amorphophallus campanulatus	Araceae	Petiole sap fermented, drunk	Mt Hagen
Areca catechu	Palmae	Quid, mixed with coconut, chewed and swallowed	New Britain
Aristolochia sp.	Aristolochiaceae	Boiled leaf extract drunk	d'Entrecasteaux
Artocarpus altilis	Moraceae	Young shoot eaten	New Britain
		Latex diluted, drunk	Manus
Barringtonia sp.	Barringtoniaceae	Apical leaves squeezed into water, drunk	d'Entrecasteaux
Boehmeria sp.	Urticaceae	Leaves rubbed on stomach	d'Entrecasteaux
Breynia cernua	Euphorbiaceae	Bark extract drunk	—
Calamus sp.	Palmae	Young shoots eaten	Widespread

Plant name	Family	Part used and treatment	Area recorded
Carica papaya	Caricaceae	Latex	Northern D., d'Entrecasteaux
Casuarina equisetifolia	Casuarinaceae	Roots washed, scraped, mixed with water, drunk	Central D.
Centella asiatica	Apiaceae	Leaves crushed, eaten or extract drunk	Trobriands
Cocos nucifera	Palmae	Roots eaten	New Britain
Codiaeum variegatum	Euphorbiaceae	Root masticated with betel nut	Manus
Colocasia esculenta	Araceae	Leaves cooked, eaten	New Britain
Cordyline fruticosa	Liliaceae	Leaves, petioles crushed, diluted, drunk	Sepik
Curcuma sp.	Zingiberaceae	Root masticated with ash-salt	Kukukuku
Dodonaea viscosa	Sapindaceae	Leaves heated, pressed against abdomen	Mt Hagen
		Inside bark crushed, diluted, extract drunk	SHD
Drimys piperita	Winteraceae	Young stems split, added to sweet potato, eaten	Tari
Elaeocarpus sp.	Elaeocarpaceae	—	—
Elatostema sp.	Urticaceae	—	Northern D.
Ficus septica	Moraceae	—	Northern D.
Ficus sp.	Moraceae	Latex mixed with *Syzygium* sp., extract drunk	—
Flagellaria indica	Flagellariaceae	Stem crushed, diluted, extract drunk	—
Hibiscus manihot	Malvaceae	Eaten with banana and *Phyllanthus* sp.	Central D.
Hornstedtia sp.	Zingiberaceae	—	—
Imperata sp.	Poaceae	Shoots chewed, sap taken	Central D.
Ipomoea pes-caprae	Convolvulaceae	Leaves masticated	Manus
Laportea decumana	Urticaceae	Leaf rubbed on stomach	SHD, Northern D.
Lophopyxis maingayi	Icacinaceae	Stem masticated	Manus
Mallotus philippinensis	Euphorbiaceae	Boiled leaf solution taken	Central D.
Mallotus ricinoides	Euphorbiaceae	Sap mixed with coconut milk, drunk	—
Morinda citrifolia	Rubiaceae	Apical leaves crushed, diluted, drunk	d'Entrecasteaux
Mucuna sp.	Leguminosae	Stem crushed, diluted extract drunk	Sepik
Musa spp.	Musaceae	Leaves, shoots crushed, diluted, soup drunk	Ramu
Nicotiana tabacum	Solanaceae	Young leaves masticated and swallowed	Sepik
Parsonsia sp.	Apocynaceae	—	Northern D.
Piper sp.	Piperaceae	—	—
Pongamia pinnata	Leguminosae	Apical leaves crushed with salt water, extract drunk	Manus
Psidium guajava	Myrtaceae	Boiled leaf solution drunk	Central D., d'Entrecasteaux
Pterocarpus indicus	Leguminosae	—	Northern D.
Quassia sp. (syn. *Simarouba*)	Simaroubaceae	—	—
Rhyticaryum sp.	Icacinaceae	Leaves eaten with traditional salt	Watut, EHD
Saccharum officinarum	Poaceae	Sugarcane eaten	New Britain
Scaevola sp.	Goodeniaceae	—	—
Schefflera sp.	Araliaceae	Leaves eaten	Watut, EHD
Sida acuta	Malvaceae	Boiled leaf solution drunk	Central D.
Sida acuta ssp. *carpinifolia*	Malvaceae	Root crushed, diluted, drunk	d'Entrecasteaux
Sida cordifolia	Malvaceae	Leaves chewed, boiled leaf solution drunk	Central D.
Symbegonia sp.	Begoniaceae	Leaves mixed with other greens, boiled, eaten with sweet potato	SHD
Syzygium sp.	Myrtaceae	Leaves boiled, eaten	New Britain
Tournefortia sarmentosa	Boraginaceae	Leaves eaten	Sepik, Central D.
Trimenia sp.	Trimeniaceae	—	—
Wedelia biflora	Asteraceae	Stem crushed, diluted extract drunk	Manus
Zingiber spp.	Zingiberaceae	Rhizome mixed with *Alstonia scholaris* extract, drunk	New Britain
		Rhizome chewed	Widespread

EHD = Eastern Highlands District
SHD = Southern Highlands District

frequently used: betel nut (*Areca catechu*) mixed with coconut is chewed and swallowed, the roots of the coconut palm itself may be eaten or the young shoots of a rattan, *Calamus* sp. Gingers are commonly used: *Hornstedtia* sp. seeds are eaten, *Curcuma* sp. and *Zingiber* spp. rhizomes chewed with ash-salt and the extract of the crushed rhizome of *Zingiber* sp., mixed with shredded bark of *Alstonia scholaris*, drunk.

Other stomach complaints. To cure constipation mucilaginous extracts of the leaves or shoots of *Hibiscus manihot* and of the bark of *Pipturus* sp. are taken and leaves of *Myristica* sp. may be eaten. Plants noted as purgatives include an unidentified Commelinacea, *Pterocarpus indicus* and *Macaranga aleuritoides*. To treat colic, worms or a bloated abdomen the stems of wild taros, leaf extracts of *Premna integrifolia* or *Aristolochia* sp. and concoctions of *Erigeron linifolius, Heteromorpha arborescens, Macaranga* sp. and *Vernonia* sp. are used. Vomiting can be stopped by eating *Desmodium ormocarpoides, Peperomia* sp. and *Zingiber* sp.

Poisons. Acute stomach pains are often considered to be the result of ritual poisoning and in such cases extracts of 'stronger' plants will be employed as antidotes and emetics, namely *Derris* sp., *Erythrina* sp., *Crinum macranthum, Ficus septica* if used with *Maesa edulis*, and, taken with ginger and ash-salt, *Harmsiopanax harmsii, Euphorbia buxoides* or *Oenanthe javanica*. The overpowering scent of some *Rhododendron* spp. flowers causes vomiting, useful as a poison antidote in some parts of the highlands. While *Alstonia scholaris, A. spectabilis* and *Euphorbia buxoides* products can be used as emetics, they are poisonous if large doses are administered. Thus they can be used as ritual poisons together with others recorded, such as *Piptadenia novoguineensis* (*prosopis insularum*) and *Dysoxylum* sp. (usually mixed together), seeds of *Pipturus* sp., *Barringtonia* sp., *Cycas circinalis*, fruits of *Drymaria cordata* and any part of *Ternstroemia* sp.

Plants associated with control of fertility and aids to childbirth. In most areas oral contraceptives, abortifacients and plants causing sterility are known. The information is closely guarded by the women, however, and their use clothed in ritual and magic. The species listed in Table 3.13 are therefore only a few of those likely to be employed. Extracts of the leaves or stems of *Callicarpa* spp., *Hemigraphis* sp., *Ludwigia adscendens* and *Scaevola taccada* are taken as oral contraceptives, while infertility may be induced by eating the leaves of *Caldesia parnassifolia*, or the stems of *Flagellaria indica*. The use of *Kaempferia galanga, Laportea decumana* and certain species of bananas are recorded as abortifacients but no details of their use or action are available (Stopp 1963; Womersley 1973; Holdsworth and Heers 1971; Holdsworth and Tringen 1973; Holdsworth and N'Drawii 1973; Webb 1960).

Miscellaneous. Internal haemorrhages are treated by drinking a boiled leaf solution of *Triumfetta* sp. or plant of *Euphorbia* sp. in the Central District. Extracts of the leaves of two legumes, *Pongamia pinnata* and *Samanea saman* are given to children to quieten them, and leaves of *Acalypha wilkesiana* are taken also as a sedative. In both the Central District and the d'Entrecasteaux Islands chewing

Table 3.13 Plants associated with control of fertility and childbirth

Plant name	Family	Use	Area recorded
Buchnera ciliata	Scrophulariaceae	Raw plant eaten to alleviate birth pains	Mt Hagen
Caldesia parnassifolia	Alismataceae	Leaf eaten, causes sterility	Mt Hagen
Callicarpa cana	Verbenaceae	Contraceptive	—
Callicarpa sp.	Verbenaceae	Contraceptive	—
Cleome viscosa	Capparidaceae	Leaves chewed with betel nut to aid conception	Central D.
Codiaeum variegatum	Euphorbiaceae	Contraceptive, leaves eaten	Kukukuku
Coleus sp.	Lamiaceae	Leaves rubbed on stomach to relieve birth pains; used with *Impatients hawkeri*	Kukukuku
Cyathula prostrata	Amaranthaceae	Terminates pregnancy	Sepik
Ficus spp.	Moraceae	Latex applied to dry up umbilicus after birth	—
Flagellaria indica	Flagellariaceae	Stem eaten, causes sterility	—
Hemigraphis sp.	Acanthaceae	Contraceptive, extract of crushed leaves taken	Trobriands
Hibiscus manihot	Malvaceae	Leaves, shoots eaten	New Britain
Hibiscus rosa-sinensis	Malvaceae	Relieves labour pains	Northern D.
Ipomoea sp.	Convolvulaceae	Root cooked with bark, eaten, induces lactation without parturition	—
Kaempferia galanga	Zingiberaceae	Abortifacient	Chimbu
Laportea decumana	Urticaceae	Abortifacient	SHD
Ludwigia adscendens	Onagraceae	Contraceptive, extract of leaf, stem taken	Sepik
Musa sp.	Musaceae	Leaves eaten, cause abortion	Sepik
Oxalis magellanica	Oxalidaceae	Increases fertility, plant eaten	Mt Hagen
Phaius tankervilliae	Orchidaceae	Flower heaten in smoke, eaten to aid conception	SHD
Pipturus argenteus	Urticaceae	Extract of crushed bark drunk, aids birth	New Britain
Polygala paniculata	Polygalaceae	Flowers eaten to combat infertility	Mt Hagen
Pueraria phaseoloides	Leguminosae	Extract of crushed bark taken, aids birth	New Britain
Scaevola taccada (syn. *S. frutescens*)	Goodeniaceae	Contraceptive, crushed leaf extract or salt water-soaked leaves taken	Manus, Trobriands
Sida rhombifolia	Malvaceae	Eases childbirth	Northern D.

SHD = Southern Highlands District

of betel nut is also recorded as being used as a sedative by Holdsworth (1974a,b). *Astelia papuana, Polyscias scutellaria, Piper* sp., *Aristolochia* sp., *Acorus calamus, Mallotus philippinensis, Rubus moluccanus* and *R. ledermannii* and *Tetracera* sp. are recorded as tonics, while various species are used as stimulants: *Pteridium aquilinum, Drymaria cordata* (eaten with ash-salt), *Alstonia scholaris* and *Arenga* sp. The use of tobacco leaves rubbed into the hair to remove lice is recorded from the Mendi area of the Highlands, and elsewhere *Castanopsis acuminatissima* is used as a vermifuge. Numerous other taxa with medicinal uses are recorded in the literature but are either unidentified or identified only to family level, or the use is not specified in detail (Hide 1974; Rappaport 1967; Stopp 1963; Straatmans 1971; Womersley 1973).

Ritual and magic

Plants play an important part in religion and magic in New Guinea. They are often used as totems in coastal areas and, throughout the country, in rituals associated with all economic, social and political aspects of life: to ensure success in crop production, pig raising, fishing, hunting, trading and warfare, to maintain health, attract a marriage partner and wealth, to become a leader (Lawrence 1972). Many species are noted in ethnographic accounts, some 130 taxa are listed in Table 3.14.

Table 3.14 Plants used in rituals and magic

Plant name	Family	Ritual	Area recorded
Acalypha sp.	Euphorbiaceae	Pig	Jimi
Acorus calamus	Araceae	Various	Widespread
Adenostemma hirsutum	Asteraceae	Pig	Highlands
Aglaonema sp.	Araceae	Dead	Jimi
Albizia sp.	Leguminosae	Fighting	Jimi
Alocasia hollrungii	Araceae	—	Chimbu
Alphitonia incana	Rhamnaceae	Dead, fighting	Mt Hagen, Jimi
Alpinia spp.	Zingiberaceae	Various	Widespread
Amaracarpus brassii	Rubiaceae	—	Chimbu
Amomum polycarpum	Zingiberaceae	Boundary protection	Jimi
Angiopteris sp.	Marattiaceae	Various	Jimi
Araucaria hunsteinii	Araucariaceae	House protection	Jimi
Ardisia sp.	Myrsinaceae	—	Chimbu
Areca catechu	Palmae	Various	Widespread coastal
Arrhenechthites sp.	Asteraceae	—	Chimbu
Artocarpus altilis	Moraceae	Rain magic	Lowlands
Astilbe sp.	Saxifragaceae	—	Chimbu
Astronia sp.	Melastomataceae	—	Jimi
Bambusa spp.	Poaceae	Various	Widespread
Beilschmiedia sp.	Lauraceae	—	Jimi, Tari
Blumea riparia	Asteraceae	—	Chimbu
Borreria laevis	Rubiaceae	—	Chimbu
Breynia sp.	Euphorbiaceae	—	Jimi
Buchnera tomentosa	Scrophulariaceae	—	Chimbu
Calanthe sp.	Orchidaceae	—	Jimi
Canavalia obtusifolia	Leguminosae	Fishing	New Britain
Canarium sp.	Burseraceae	Garden magic	Coastal (Rai)
Capillipedium parviflorum	Poaceae	—	Chimbu
Carex sp.	Cyperaceae	—	Chimbu
Carpodetus sp.	Saxifragaceae	—	Chimbu
Casuarina papuana	Casuarinaceae	Dead	Tari
Casuarina spp.	Casuarinaceae	Garden	Widespread
Chloranthus sp.	Chloranthaceae	—	Jimi
Cinnamomum sp.	Lauraceae	—	Chimbu
Claoxylon sp.	Euphorbiaceae	—	Chimbu
Cleistanthus sp.	Euphorbiaceae	House	Jimi
Cocos nucifera	Palmae	Various	Lowlands
Codiaeum variegatum	Euphorbiaceae	Boundary protection	Coastal, highlands
Coleus atropurpureus	Lamiaceae	Rain magic	New Britain
Coleus scutellarioides	Lamiaceae	Various	Widespread
Colocasia esculenta	Araceae	Garden	Widespread
Colona scabra	Tiliaceae	Fighting	Jimi
Cordyline fruticosa	Liliaceae	Garden, various	Widespread
Crinum sp.	Amaryllidaceae	Medical, various	Enga, Chimbu
Crotalaria ferruginea	Leguminosae	—	Chimbu
Croton sp.	Euphorbiaceae	Garden	Widespread
Cryptocarya sp.	Lauraceae	Fighting	Jimi
Cymbopogon flexuosus	Poaceae	Garden, various	Widespread
Cynoglossum helwigii	Boraginaceae	—	Chimbu
Cyrtandra sp.	Gesneriaceae	—	Jimi, Chimbu
Decaspermum sp.	Myrtaceae	Medical, various	Enga, Chimbu
Desmodium spp.	Leguminosae	—	Chimbu
Dioscorea spp.	Dioscoreaceae	Garden, various	Widespread
Dodonaea viscosa	Sapindaceae	Dead	Baliem
Elaeocarpus spp.	Elaeocarpaceae	—	Chimbu
Elatostema blechnoides	Urticaceae	Pig feast	Jimi, Tari
Eleusine indica	Poaceae	—	Chimbu
Elmerrillia papuana	Magnoliaceae	—	Jimi
Endospermum formicarum	Euphorbiaceae	Fighting, love magic	Jimi, New Britain

Plant name	Family	Ritual	Area recorded
Euodia spp.	Rutaceae	Dead	Tari
Euphorbia buxoides	Euphorbiaceae	Medical	—
Evodiella hooglandii	Rutaceae	Love magic	SHD, Chimbu
Ficus spp.	Moraceae	Rain magic, combat sorcery	Widespread
Galbulimima belgraveana	Himantandraceae	Fighting	—
Garcina spp.	Clusiaceae	Medical, various	Highlands
Grevillea papuana	Proteaceae	—	Chimbu
Hemigraphis sp.	Acanthaceae	Love magic	Enga
Hibiscus sp.	Malvaceae	Garden	Trobriands
Holochlamys sp.	Araceae	Medical, with *Euodia* sp.	Tari
Ischaemum barbatum	Poaceae	—	Chimbu
Litsea sp.	Lauraceae	—	Jimi
Lycianthes sp.	Solanaceae	—	Jimi
Lycopodium sp.	Lycopodiaceae	Medical	Hagen, Jimi
Mearnsia cordata	Myrtaceae	Women's rites	SHD
Medinilla sp.	Melastomataceae	—	Chimbu
Muehlenbeckia platyclada	Polygonaceae	—	Chimbu
Musa spp.	Musaceae	Various	Widespread
Myrmecodia sp.	Rubiaceae	—	Chimbu
Nephrolepis sp.	Oleandraceae	Dead	Tari
Nertera granadensis	Rubiaceae	—	Chimbu
Nicotiana tabacum	Solanaceae	Various	Widespread
Ocimum basilicum	Lamiaceae	Rain magic	New Britain
Octomeles sumatrana	Tetramelaceae	—	New Britain
Omalanthus sp.	Euphorbiaceae	Fighting	New Britain
Oxalis sp.	Oxalidaceae	Protection against sorcery	d'Entrecasteaux
Pandanus spp.	Pandanaceae	Medical, initiation, dead	Enga, Kukukuku, Tari
Pangium edule	Flacourtiaceae	—	Jimi
Pilea sp.	Urticaceae	Initiation, various	Kukukuku, Chimbu
Piper sp.	Piperaceae	Various	Widespread
Pipturus sp.	Urticaceae	Dead, various	Jimi, Chimbu
Podocarpus sp.	Podocarpaceae	Various	Tari
Polygonum chinense	Polygonaceae	Pig	Tari
Polygonum minus	Polygonaceae	Sickness	Tari
Polypodium spp.	Polypodiaceae	Dead	Tari
Psychotria spp.	Rubiaceae	Various	Chimbu
Pteridium aquilinum	Dennstaedtiaceae	—	Chimbu
Ranunculus pseudolowii	Ranunculaceae	—	Chimbu
Rapanea sp.	Myrsinaceae	House, sickness	Tari
Rhamnus nepalensis	Rhamnaceae	—	Chimbu
Rhododendron spp.	Ericaceae	Various	Chimbu
Rubus rosifolius	Rosaceae	Medical	Tari
Saccharum officinarum	Poaceae	Various	Widespread
Solanum torvoideum	Solanaceae	Fighting	Tari
Sopubia trifida	Scrophulariaceae	—	Chimbu
Spathoglottis sp.	Orchidaceae	—	Jimi
Spondias dulcis	Anacardiaceae	—	Jimi
Stellaria spp.	Caryophyllaceae	—	Chimbu
Syzygium sp.	Myrtaceae	—	Chimbu, Jimi
Trema sp.	Ulmaceae	Fishing	New Britain
Trimenia sp.	Trimeniaceae	—	Chimbu
Triumfetta rhomboidea	Tiliaceae	—	Chimbu
Vaccinium acrobracteatum	Ericaceae	—	Chimbu
Vernonia lanceolata	Asteraceae	—	Chimbu
Wahlenbergia marginata	Campanulaceae	—	Chimbu
Zingiber spp.	Zingiberaceae	Various	Widespread

SHD = Southern Highlands District

The cultivated plants are of particular importance: food plants such as coconuts, yams, taro, bananas, sugarcane, condiments such as ginger and *Piper* spp., stimulants such as areca nut and tobacco and the ornamental species, *Cordyline fruticosa, Codiaeum variegatum, Coleus scutellarioides* and *Acorus calamus* among others. Species used for canoes (*Canarium* spp., *Octomeles sumatrana*) are important in coastal areas. Other species are gathered from forest and grassland when required. Ginger and sugarcane and *Acorus calamus* in some areas are extremely important in ritual medicine and in counteracting poison and sorcery. *Cymbopogon* sp., *Euphorbia buxoides* and *Garcinia* sp. are used ritually to prevent transference of sickness from one person to another, while *Pandanus* sp., *Decaspermum* sp. and *Crinum* sp. may be planted to help strengthen a sick person. Abortifacient techniques, although having a pragmatic base (Table 3.13) are regarded as a form of ritual, usually performed by women. Species used in rituals associated with the dead include *Dodonaea viscosa*, used to anoint the dead body and as incense in cremation fires in the Baliem valley, *Euodia* spp. and ferns (*Nephrolepis* sp., *Polypodium* spp.) in the Tari area. The use of bark, bark cloth, mats, sago and banana leaves for wrapping the corpse is widespread.

Acorus calamus is important in many areas, being used in diverse rituals, to make young men grow tall and strong, to promote success in hunting, to attract wealth and to prevent face paint from running during ceremonial dancing. Other species used to make young warriors fierce include *Omalanthus* sp., *Zingiber* sp., *Endospermum formicarum,* and *Galbulimima belgraveana*. Tobacco, bamboo, *Cordyline fruticosa, Acorus calamus, Hemigraphis* sp., *Evodiella hooglandii, Cinnamomum* spp. and *Endospermum formicarum* as well as many scented plants (Chowning, pers. comm.) are used in love magic.

Garden magic employs, as well as cultivated food species, *Coleus* spp., *Cymbopogon flexuosus, Casuarina* sp., *Hibiscus* sp., *Croton* sp. leaves and those of wild fruit and nut trees, among others. *Cordyline fruticosa, Codiaeum variegatum* and *Amomum polycarpum* are recorded widely as used for boundary protection. Banana leaves, sago, *Ocimum basilicum, Coleus atropurpureus,* coconut milk, the latex of *Ficus* spp., *Artocarpus altilis,* and sap of other trees are important in rain magic. Cooked leaves of *Canavalia obtusifolia* are rubbed on fishing lines to bring good luck, and the bark of *Trema* sp. used to hold odoriferous plants used in fishing magic (Panoff 1972; Feachem 1973; Floyd 1954; Lawrence 1972; Malinowski 1935).

Art

Masks, carvings and paintings are associated with religion and magical rites throughout coastal and lowland regions. The types of art present and the distribution of art styles are well documented (Newton 1972). A fairly wide range of plant materials is recorded as used but no detailed study has been made of them. *Vitex cofassus, Intsia bijuga* and *Albizia* sp. woods and many species of Palmae are widely used for carvings. Leaves of *Ficus* spp. or

Equisetum debile stems are used as abrasive sandpaper to smooth surfaces. Paintings are done on sago and other palm spathes and on bark cloth made from *Ficus* spp., *Broussonetia papyrifera* and *Hibiscus* sp. among others. A wash of *Artocarpus altilis* sap may be used to prepare the bark for paint in some areas. Masks are made from wood, tree fern roots, bark cloth stretched on a wooden frame, palm leaves and spathes, coconut shells or gourds, and woven *Lygodium* fern stems or rattan cane (*Calamus* sp.). Skirts for masks are of shredded sago or coconut palm leaf or banana fibre, and other decorations may incorperate *Coix lachryma-jobi* seeds or *Abrus* sp. kernels (Cranstone 1961). Figureheads for daggers and combs may be modelled from *Maranthes* nut paste. Paints and dyes are obtained from the roots of *Morinda citrifolia* and *Curcuma domestica,* fruits of *Burckella* sp., *Leea indica* and *Pandanus conoideus,* the pulp surrounding the seeds of *Bixa orellana,* seeds of *Areca* spp., *Pittosporum pullifolium, Gardenia* sp. and *Lactuca indica,* crushed leaves and stems of *Plectranthus* sp., *Iresine herbstii* and *Coleus* spp., petals of *Bidens* sp. and *Tagetes* sp., and *Melastoma malabathricum.* Red paint is made from the bark of *Glochidion* sp. and from the sap of an *Acalypha* sp. The sap of Rhizophoraceae, coconut oil or oil from the seeds of *Aleurites moluccana* is used as a base for the paints and for charcoal. Paintbrushes are made of coconut or *Pandanus* spp. fibre, the husk of *Areca* nuts, the midrib of a *Dracontomelon puberulum* (*Dracontomelum mangiferum*) leaf or a sliver of bamboo. *Cycas* sp. is used as a paint container in the Wissel Lakes area (Pospisil 1963). Soaking of materials in organic muds and peats results in grey or black colouration (A. Bühler 1948; Floyd 1954; Held 1957, Williams 1930, 1936; Blackwood 1940).

Musical instruments comprise mainly drums, pan-pipes, flutes and the Jew's harp. Bullroarers, used in rituals, are made from palm wood or a fern, *Cyathea tenggerensis* (*Alsophila glauca*) (Williams 1930; Blackwood 1940); trumpets of bamboo, wood or tree fern stems are also recorded from the Orokaiva area and elsewhere (Chowning, pers. comm.). Drums are used widely; they may be 30-200 cm long and are made from hollowed lengths of *Hernandia nymphaeaefolia* (*H. peltata*), *Pterocarpus indicus, Premna integrifolia, Araucaria* sp.) and occasionally from bamboo (Floyd 1954; Landtman 1933). The sap of *Rhododendron* sp., *Dimorphanthera* sp., *Artocarpus altilis, Pipturus* sp. or *Ficus* spp., the resin of *Araucaria* sp., or even starchy, cooked taro is used to fasten the drum skin to the head on kundu drums, or the skin may be bound on to the head with specially prepared rope, banana fibre or rattan cane or the bark of *Pipturus* sp. Garamut drums are often made from *Albizia* sp. or *Intsia bijuga* wood, and decorated with leaves of *Desmodium umbellatum*; *Calamus* sp. stems are used to beat the drum (Floyd 1954). For the Chimbu area of the Highlands Hide (1974) records the use of *Cerbera* sp., *Meliosma* sp., *Aleurites* sp. and *Symplocos* sp. for musical instruments. Pan-pipes made from four or five sections of different length, 1-2 cm diameter bamboo tied together are widely used. Flutes and the Jew's harp are also made from bamboo. A whistle made from a small coconut shell with a mouth-hole and two small stops is recorded by Williams (1936) for the Trans-Fly area; elsewhere whistles of bamboo are occasionally used. Seeds of *Crotalaria ferruginea* and a number of other legumes are used in rattles.

Bows may be used as a simple stringed instrument, the bowstring being tapped with an arrow or played with the mouth (Landtman 1933; Cranstone 1972). Other stringed instruments are made with sago palm leaf midribs, or bamboo (Williams 1930, 1936).

Tools and weapons

Table 3.15 lists the species recorded as used in making tools and weapons throughout New Guinea. Axes and adzes are used for all woodworking: felling trees, shaping house and canoe planks, spear shafts, hollowing out canoes, food bowls, and so on. The stone blade is hafted in various ways. It may be inserted directly into a hole in the haft made from a tree root or lower stem and root of bamboo, as in southern parts of Irian Jaya and the Trans-Fly; lashed to an 'elbow' made from a tree branch of *Bruguiera gymnorhiza* or *Piptadenia novoguineensis* (*Prosopis insularum*) among others; or fastened into a socket that is lashed to a handle. Recorded as used for the third method are species of *Decaspermum, Elaeocarpus,*

Table 3.15 Plants used to make tools and weapons

Plant name	Family	Use	Area recorded
Acalypha sp.	Euphorbiaceae	—	Chimbu
Acronychia sp.	Rutaceae	Digging sticks	Mt Hagen
Albizia falcataria	Leguminosae	Shields	Jimi
Alectryon sp.	Sapindaceae	Toy bows and arrows	Tari
Araucaria cunninghamii	Araucariaceae	Resin used as glue for stone club head, resin burnt as torch	Kukukuku
Ardisia sp.	Myrsinaceae	—	Chimbu
Areca catechu	Palmae	Husk as paintbrush	Orokaiva
Areca sp.	Palmae	Bows	Kukukuku
Bambusa spp.	Poaceae	Knives, bows, arrow heads, axe hafts, bowstring, spear thrower, pen and paintbrush, 'armour', torch holder	Widespread
Breynia sp.	Euphorbiaceae	Digging sticks	Tari
Bridelia sp.	Euphorbiaceae	Axe handles	New Britain
Bruguiera gymnorhiza	Rhizophoraceae	Axe-haft, digging sticks	New Britain, Trobriands
Calamus spp.	Palmae	Binding on arrows, adzes; bowstrings, bow bracer, stop for bowstring	Widespread
Calophyllum inophyllum	Clusiaceae	Fruit latex as glue	New Britain
Calophyllum sp.	Clusiaceae	Spear	New Britain
Carpodetus sp.	Saxifragaceae	—	Chimbu
Caryota sp.	Palmae	Bows	Idenburg R.
Castanopsis acuminatissima	Fagaceae	Shield	Mt Hagen
Castanospermum australe	Leguminosae	Spear	New Britain
Casuarina spp.	Casuarinaceae	Arrow heads, digging sticks	Wissel Lakes, Mt Hagen, Chimbu
Claoxylon sp.	Euphorbiaceae	—	Chimbu
Cocos nucifera	Palmae	Fibre as paintbrush, shell for scoops, spoons, knife to cut and peel taro	Trans-Fly, Waropen, Orokaiva
Cyathea sp.	Cyatheaceae	Fishing spear	New Britain
Decaspermum sp.	Myrtaceae	Axe socket	New Britain
Dicranopteris linearis	Gleicheniaceae	Core fibre for bindings	Jimi
Diospyros sp.	Ebenaceae	Digging stick	New Britain

Plant name	Family	Use	Area recorded
Dodonaea viscosa	Sapindaceae	Digging stick, various	Mt Hagen, Chimbu, Tari
Dracontomelon puberulum (syn. *Dracontomelum mangiferum*)	Anacardiaceae	Midrib of leaf as paintbrush	New Britain
Elaeocarpus sp.	Elaeocarpaceae	Axe handles	Wissel Lakes
Erythrina orientalis	Leguminosae	Spear, seeds as decoration on stone club head	New Britain, Kukukuku
Equisetum debile (syn. *E. indica*)	Equisetaceae	File, sandpaper	Tari, Chimbu
Euodia sp.	Rutaceae	Digging sticks, glue for bindings	Mt Hagen
Euroschinus papuanus	Anacardiaceae	Latex as glue	Jimi
Euphorbia buxoides	Euphorbiaceae	Latex as glue for bindings	Mt Hagen
Evodiella sp.	Rutaceae	Glue for bindings	Mt Hagen
Fagraea ceilanica	Loganiaceae	Timber for weapons	Chimbu
Fagraea sp.	Loganiaceae	Timber for spade handles	Tari
Ficus copiosa	Moraceae	Club for gardening	Tari
Ficus gul	Moraceae	Glue for bindings	Mt Hagen, Tari
Ficus trachypison	Moraceae	Leaf as sandpaper	Jimi
Ficus spp.	Moraceae	Leaves as sandpaper, clubs for taro gardening, digging sticks, spade handles	Mt Hagen, Tari
Garcinia spp.	Clusiaceae	Axe handles, axe haft	Mt Hagen, Tari, Jimi, Kukukuku
Gardenia sp.	Rubiaceae	Digging sticks	Tari
Glochidion spp.	Euphorbiaceae	—	Chimbu
Gronophyllum chaunostachys	Palmae	Arrow foreshafts	Jimi
Homalium foetidum	Flacourtiaceae	Digging sticks	New Britain
Hydnocarpus sp.	Flacourtiaceae	Digging sticks	Mt Hagen
Ilex sp.	Aquifoliaceae	Arrowheads	Wissel Lakes
Licuala sp.	Palmae	Bows, arrow points, foreshafts	Jimi
Maniltoa sp.	Leguminosae	Axe handle	New Britain
Memecylon sp.	Melastomataceae	Stick to hit children	Tari
Metroxylon sp.	Palmae	Spathe, bark as bow bracer	Fly R.
Miscanthus floridulus	Poaceae	Arrow shafts	Highlands
Mussaenda sp.	Rubiaceae	Bows	Wissel Lakes
Neonauclea sp.	Rubiaceae	Digging sticks	Mt Hagen
Nothofagus sp.	Fagaceae	Digging sticks, spears	Mt Hagen
Octamyrtus pleiopetala	Myrtaceae	Axe handles	Tari
Olearia sp.	Asteraceae	—	Chimbu
Oncosperma tigillaria	Palmae	Arrow heads	Wissel Lakes
Oncosperma sp.	Palmae	Arrow heads	Kukukuku
Orania sp.	Palmae	Arrow foreshafts	Jimi
Pandanus spp.	Pandanaceae	Arrow heads, fibre as paintbrush	Wissel Lakes
Pennisetum macrostachyum	Poaceae	Arrow shafts	Chimbu
Phyllanthus nervosus	Euphorbiaceae	Digging sticks	Mt Hagen
Piper sp.	Piperaceae	Wood burnt as torch	New Britain
Piptadenia novoguineensis (syn. *Prosopis insularum*)	Leguminosae	Axe haft	New Britain
Podocarpus spp.	Podocarpaceae	Spears	Mt Hagen
Pullea glabra	Cunoniaceae	—	Chimbu
Quintinia sp.	Saxifragaccac	Digging sticks	Tari
Rapanea sp.	Myrsinaceae	Digging sticks, weapons	Tari, Chimbu
Rhus taitensis	Anacardiaceae	Glue on bindings	Mt Hagen
Rhus sp.	Anacardiaceae	Gum	Chimbu
Saccharum sp.	Poaceae	Arrow shafts	Fly R., Kukukuku
Scleria chinensis	Cyperaceae	Leaf as knife	New Britain
Selaginella caudata	Selaginellaceae	Whole plant as broom	New Britain
Symplocos sp.	Symplocaceae	Weapons	Chimbu
Syzygium spp.	Myrtaceae	Axe handles, digging sticks	Tari, New Britain
Typha sp.	Typhaceae	Torch	Tari
Vitex cofassus	Verbenaceae	Axe handles	New Britain
Xanthomyrtus sp.	Myrtaceae	Digging stick, harvesting stick	Tari, Wissel Lakes

Garcinia, Syzygium and *Octamyrtus pleiopetala* in the highlands, and *Vitex cofassus, Maniltoa* sp. and *Bridelia* sp. among others in coastal areas (Williams 1936; Cranstone 1961; Floyd 1954).

Knives may be made from stone or obsidian, black palm wood or shell. The everyday knife, used for butchering animals, carving meat, cutting hair, trimming bark cloth, and in surgery and dentistry in many areas, is made of bamboo. An internode of bamboo is selected and cut with an axe and then a sharp edge is produced by tearing away a fine strip with the teeth. The knife may be sharpened whenever necessary by following the same procedure. In some areas, like New Britain, bamboo knives are used only for butchering animals, while surgery is carried out with obsidian, haircutting and other general tasks with shell, and garden work with a knife of black palm (Chowning, pers. comm.). The sharp-edged leaf of the sedge *Scleria chinensis* is also used as a knife in that area (Floyd 1954).

Bamboo drills are used to make shell ornaments and to produce fire (Held 1957; Williams 1936). Fire is more often produced by the use of a fire-saw: a thong of bamboo or rattan cane is pulled quickly back and forth across a piece of split soft wood (*Calophyllum* sp., *Hibiscus* sp., *Symplocos* sp.) with tinder underneath. Fire-plows are used in New Britain and in Milne Bay District (Chowning, pers. comm.).

Various types of implement are used in gardening (Nilles 1942-5; Heider 1970; Pospisil 1963; Lerche and Steensberg 1973; Powell 1974a). Large diameter pointed digging sticks (sharpened on one side or all around) are used for slashing undergrowth, as crowbars for uprooting tree stumps and cane grass clumps and for planting and harvesting by the men. They are made from any heavy, straight timber available at the garden site, often *Casuarina* spp., *Dodonaea viscosa* or *Schuurmansia henningsii* in the highlands. Smaller diameter pointed digging sticks are used by the women for planting and harvesting (and occasionally as a weapon); the species used are chosen more carefully and are often hardwoods gathered from the forest. *Diospyros* sp., *Homalium foetidum, Syzygium* sp. and mangrove species are recorded as used for digging sticks in coastal areas, *Casuarina* spp., *Euodia* sp., *Breynia* sp., *Ficus* spp., *Acronychia* sp., *Gardenia* sp., *Rapanea* sp., *Nothofagus* sp., *Phyllanthus nervosus, Quintinia* sp. and *Xanthomyrtus* spp. from highland areas. Paddle-shaped spades used for drain-digging are recorded also from various parts of the highlands. They are usually made from *Casuarina* spp., but may be of *Neonauclea* sp. or *Hydnocarpus* sp. wood. A smaller hastate-shaped spade of *Nothofagus* sp., recorded from the Mt Hagen area, is used for marking plot lines and making holes for taro plants. A short club is often used for breaking up clods of earth in garden beds, or a longer one for making taro holes in the Tari area; both types are made from *Ficus copiosa* and other *Ficus* spp.

Spears and bows and arrows are used as the main weapons of warfare and also in hunting and fishing. One-piece hardwood spears with plain or barbed heads are common and are used for both thrusting and throwing. Black palm, *Calophyllum* sp., *Castano-*

spermum australe and *Erythrina orientalis* (*E. indica*) saplings are used in coastal New Britain for spears (Floyd 1954); elsewhere black palm wood or fine-grained woods such as *Nothofagus* sp. and *Podocarpus* spp. are used. Throwing spears may have bamboo or cane shafts and heads of hardwood, bone, or sting-ray spines. Fishing spears have multiple wood points attached to a bamboo shaft or may be fashioned from the trunk of a tree fern, *Cyathea* sp. A spear-thrower of bamboo is recorded for the lower Sepik-Ramu area (Cranstone 1961).

Bows are made from bamboo with strings of split rattan or bamboo in southern parts of New Guinea and some parts of the highlands; elsewhere they are usually made from black palm wood (*Caryota* sp., *Areca* sp., others unidentified), hardwoods, *Mussaenda* sp. and a Sapindacea (Pospisil 1963), or mangrove species. The bowstring may be frayed out at both ends for tying (Landtman 1933), but usually loops are tied at the ends and the bowstring held in place on the bow with plaited rattan rings. Bow bracers are made from plaited rattan cane, sago and other palm spathes or slats of sago bark bound with cane (Williams 1936).

Arrows usually have light cane shafts of *Saccharum* sp. or *Miscanthus floridulus* with foreshafts of palm wood such as *Orania* sp. and *Gronophyllum chaunostachys* and heads of various design (plain or barbed, sharp or blunt, single or multiple points) also made from palms (*Oncosperma* spp.) or other woods (*Ilex* sp., *Casuarina* spp.), *Pandanus* spp. or bamboo. The common, lanceolate-shaped, sharp-pointed hunting arrow used for killing pigs, cassowaries and wallabies is made from *Bambusa* spp.

Calamus spp. are used for bindings on all spears, bows and arrows, and the sap of various species, including *Euroschinus papuanus, Euphorbia buxoides, Calophyllum inophyllum, Euodia* spp., *Evodiella* sp., *Araucaria cunninghamii* and *Rhus taitensis* (mixed together) and *Ficus gul* used as glue on arrow bindings.

Clubs are used in all areas; they are of various designs and made from palm wood or from natural swellings of tree branches and roots. Others have fitted stone heads. Species of Rubiaceae, Myrsinaceae and Moraceae are the materials used in the highlands for clubs (Blackwood 1950). Seeds of *Erythrina* sp. may be set in resin around a stone club head as decoration in the Kukukuku area.

Cuirasses of plaited rattan are recorded from western parts of New Guinea and south-western Papua New Guinea (Riesenfeld 1946) while elsewhere broad bamboo belts and wooden plates have been noted as 'armour'. Wooden shields are quite common; they are usually rectangular in shape and carved in relief and/or painted. *Albizia falcataria* wood was used in their manufacture in the Jimi valley, *Castanopsis acuminatissima* in the Mt Hagen area.

Hunting and fishing

Hunting and trapping methods have been described in detail in Anell (1960) and Bulmer (1968). The use of spears and of bows and arrows has already been mentioned. Hand nets and longer nets are used to catch pigs and also birds and bats in some areas; they are made from fibre ropes of *Trema* sp., *Leucosyke* sp., *Hibiscus tiliaceus, Abroma augusta* and *Ficus* spp. among others. A 'pig-fender', comprising a loop of *Calamus* sp. cane fastened to a pole and sometimes with an attached net, is used to catch wild animals alive. Simple hand-operated snares made from rope or rattan are common throughout New Guinea and are often baited with wild fruits, sago or bananas. Spring snares are usually made from bamboo or flexible withies of *Syzygium* spp., *Aglaia sapindina* and *Dodonaea viscosa* among others. Tube snares are uncommon; rat traps comprising a baited hollow stem of a *Piper* sp. or *Pandanus* sp. log and a bamboo tube with spring snare are recorded. Blow guns of bamboo are used for hunting in south-west New Britain (Chowning, pers. comm.).

Dead falls, log traps and spiked pit falls are used mainly to catch pigs; they include various timbers, bamboo and canes. Spikes are frequently made from bamboo, driven into the ground near fence-lines. Box traps of rattan or bamboo, usually wide at the entrance and tapering to a point, sometimes lined with thorns, are commonly used for catching smaller animals. A pole with a bundle of thorny canes (*Calamus* sp.) is swung around to catch bats or a thorn-lined conical basket is thrown over them (Williams 1930). Bird whistles made from a certain fruit or from coconut shell are also used in some lowland areas and the flattened, fleshy stem of an epiphytic orchid is recorded as a bird whistle in the Jimi valley area of the highlands (Clarke 1971).

The use of bird lime is widespread. The sticky latex of *Artocarpus altilis* (sometimes mixed with coconut milk) and of *Rhus taitensis, Ficus dammaropsis, Ficus myriocarpa* and other *Ficus* spp., the resin of *Euodia* sp. and *Evodiella* sp., and the sticky fruits of *Pisonia longirostris* in coastal New Britain (Floyd 1954), are applied to tree branches and poles.

Fishing methods are discussed in Anell (1955), Cranstone (1961, 1972) and Malinowski (1918). Multi-pointed, barbed spears and arrows are commonly used in fishing, and harpoons, used for catching turtle and dugong, are recorded from parts of Irian Jaya. Fish traps are of various kinds. Wooden stake fences or stake plus net or mat weirs are used to dam creeks or block off parts of a lagoon; simple tubular basket traps are placed in fast-flowing waters and baited conical traps with movable slat entrances or thorny linings are anchored in rivers and coastal waters. Conical plunge-baskets are widely used to catch shoal fish in shallow waters. Such traps are made from *Calamus* spp. cane, bamboo, *Lygodium* sp. fibre or, occasionally, bark of *Pandanus* spp., Lauraceae and other forest trees. Laths of the bark of sago leaf midribs are made into a mat used as a fish trap in the Waropen area of Irian Jaya (Held 1957). Saplings of *Allophylus cobbe* are used as fish trap markers in some areas (Floyd 1954). The rough leaves of *Ficus pachysrhachis* are used for seizing eels in the Jimi valley (Clarke 1971).

Dip nets of *Ficus* spp. or *Pandanus* spp. rope, with wooden (*Acalypha* sp., *Dodonaea viscosa*) or bamboo frames and handles are used in shallow lake waters and along coastal reefs (Pospisil 1963; Held 1957; Williams 1936). Seine nets and casting nets are used in eastern Papua New Guinea for catching shoal fish. Net supports may be made from *Calophyllum inophyllum.*

Fishing lines are made from coconut husk fibres and roots, fibres of aerial roots of *Pandanus* spp. or stems of bananas, or the more usual fibre trees. Hooks and gorges are made from palm wood or the thorny *Pandanus* or sago palm bark. A sago palm leaf rib acts as a float. The silvery insides of *Crinum asiaticum* leaves are used as lures when trolling for fish in the New Britain area (Floyd 1954). In the Trobriands and Admiralty Islands sharks are attracted to canoes with a decoy-rattle, consisting of coconut shell segments threaded on a bent stick, and then caught with a rope or rattan noose.

Poisons are frequently used to drug fish in rivers and shallow waters. Crushed seeds of *Barringtonia* spp. and *Cerbera odollam,* roots of *Derris* sp. and *Pongamia pinnata,* bark of *Pangium edule* and a Theacea, and the latex of *Euphorbia plumerioides* and *E. buxoides* are recorded from various areas.

Torches used for night fishing and bait catching may be made from bunches of dried coconut leaves, the prop roots of *Pandanus papuanus,* wood of *Piper* sp. or a bamboo internode filled with *Araucaria cunninghamii* resin.

Canoes and rafts

Many different types of canoes are recorded from both coastal and inland waterways (Haddon and Hornell 1937). They range from the simple dugout (Plate 48) used for everyday purposes of transport, fishing and gathering to the more complex double-outrigger sailing canoes and multi-hulled vessels involved in long distance trading voyages. Rafts are less often used in coastal areas (recorded from

Plate 48
Trunk of *Calophyllum* tree being hollowed out and made into a canoe.

New Britain, Watom Island and New Ireland) but are important in crossing lakes and rivers in the Gulf, Western and Northern Districts of Papua New Guinea. No detailed study of plants utilised in canoe and raft building has been undertaken but those listed in Table 3.16 give some indication of the wide range of species considered suitable.

Table 3.16 Plants used in the construction of canoes and rafts

Plant name	Family	Use	Area recorded
Albizia falcataria	Leguminosae	Canoe hulls	New Britain
Alstonia sp.	Apocynaceae	Canoe hulls	New Britain, New Ireland
Althoffia pleiostigma	Tiliaceae	Canoe hulls	—
Areca catechu	Palmae	Stem as wash-strake, leaf petiole as fire hearth, and platform	New Britain, Watom Is., Siassi Is., Aitape
Arenga saccharifera	Palmae	Leaf midrib fibres as lashing, decoration	Geelvink Bay
Artocarpus sp.	Moraceae	Canoe hulls	Maty Is.
Bambusa spp.	Poaceae	Platforms, decking, masts, booms, floats, spars, wash-strakes	Watom Is., Papua Gulf, Central D., Nissan Is., Geelvink Bay, Waropen, Humboldt Bay
		Raft	Trans-Fly, Kiwai, Watom Is.,
		Bailer	Nissan Is., New Ireland
Bruguiera spp.	Rhizophoraceae	Gunwales, masts, booms, spars, outrigger poles	Papuan Gulf, Central D.
Calamus spp.	Palmae	Stem for lashings, bindings, braces, anchor cable	New Ireland, Central D., Geelvink Bay
Callicarpa pentandra	Verbenaceae	Canoe stringers	New Britain
Calophyllum inophyllum	Clusiaceae	Canoe outriggers	New Britain
Calophyllum sp.	Clusiaceae	Canoe hulls	Wissel Lakes
Campnosperma sp.	Anacardiaceae	Canoe hulls	New Britain
Canarium sp.	Burseraceae	Canoe hulls	New Britain
Cinnamomum sp.	Lauraceae	Canoe hull	Northern D.
Cocos nucifera	Palmae	Plaited sails, weather screen, bailers from leaves	Aitape, Bougainville, Nissan Is., Central D.
		Fibre for lashings, decoration	New Britain, Hermit Is.
		Leaf sheath for sails	Sepik, Aitape
		Nut as bailer	Bougainville, New Britain, Geelvink Bay
Erythrina sp.	Leguminosae	Canoe hulls, floats	Fly River
Euodia elleryana	Rutaceae	Resin for caulking canoes	New Britain
Ficus myriocarpa	Moraceae	Latex for caulking canoes	New Britain
Ficus spp.	Moraceae	Canoe hulls	New Britain
Flagellaria indica	Flagellariaceae	Stem as anchor cable	New Britain
Gmelina sp.	Verbenaceae	Canoe hulls	New Britain
Hibiscus sp.	Malvaceae	Canoe hulls, outrigger floats, booms	Admiralty Is., Fly River, Louisiade Is.
Homalium foetidum	Flacourtiaceae	Canoe paddles	New Britain
Lygodium circinnatum	Schizaeceae	Lashings	New Britain, New Ireland
Macaranga aleuritoides	Euphorbiaceae	Outriggers	New Britain
Macaranga tanarius	Euphorbiaceae	Outriggers	New Britain
Macaranga sp.	Euphorbiaceae	Canoe hulls	Wissel Lakes
Mangifera sp.	Anacardiaceae	Outrigger booms	New Britain
Maranthes corymbosa (syn. *Parinari corymbosum*)	Chrysobalanaceae	Seeds for caulking canoes	Aitape, New Britain, New Ireland, Bougainville
Musa spp.	Musaceae	Stems as raft	Orokaiva, Watom Is.
		Leaves as weather screen	Central D.
Metroxylon spp.	Palmae	Stem as canoe, outrigger floats	Waropen, Frederik-Hendrik Is., Fly R., Gulf D.
		Petioles as weather screen	Geelvink Bay
		Leaves as sails, spars	Central D., Bougainville

Plant name	Family	Use	Area recorded
Nauclea coadunata (syn *N. orientalis*)	Rubiaceae	Canoes	—
Octomeles sumatrana	Tetramelaceae	Canoe hulls	New Britain
Pandanus spp.	Pandanaceae	Leaves as sails, awnings, deckhouse, bailers	Waropen, Humboldt Bay, Siassi Is., Aitape, Trobriands, Central D., New Ireland, Tari
Podocarpus sp.	Podocarpaceae	Canoe hulls	Wissel Lakes
Premna obtusifolia	Verbenaceae	Canoe nails	Gulf of Carpentaria
Thespesia populnea	Malvaceae	Canoe hulls	Admiralty Is.
Toona ciliata (syn. *Cedrela toona*)	Meliaceae	Canoe hulls	Wissel Lakes, Northern D., New Britain

Medium hard and softwood species such as *Gmelina* sp., *Canarium* sp., *Albizia falcataria, Alstonia* sp., *Campnosperma* sp. and *Octomeles sumatrana* are recorded as used for canoe hulls in New Britain (Floyd 1954) and are probably much more widely used; *Toona ciliata* (*Cedrela toona*) is important there and also in the Wissel Lakes area of Irian Jaya (Pospisil 1963), while *Gmelina moluccana* is used widely in the Solomon Islands (Whitmore 1966a). Softwoods such as *Macaranga* spp., *Hibiscus* sp., *Calophyllum* sp., and *Ficus* spp. may also be used for canoe hulls but are more often used as outrigger floats and for booms and spars. The hollowed-out stem of the sago palm (*Metroxylon* sp.) is used in the Fly River and Waropen areas where few suitable canoe timbers are available. In such areas mangrove species are used for masts, booms and spars, outrigger poles and gunwales. Bamboo is used in many areas for decking and platforms, for masts and spars, outrigger booms and floats; in others *Areca catechu* or other palms may be used, e.g. in New Britain (Chowning, pers. comm.). Bamboo bailers are recorded from New Ireland and the Nissan Islands.

For caulking planked canoes, mending leaks and patching holes the plastic putty of *Maranthes corymbosa* (*Parinari corymbosum*) seeds, the latex of *Ficus myriocarpa* and the resin of *Euodia elleryana* are used.

Plaited coconut or sago palm leaves are widely used as sails, the midrib of the leaf forming the central and lateral margins. Elsewhere *Pandanus* spp. are used for sails, the leaf midrib being removed and the margins sewn together. Leaves of *Pandanus* spp. and of the coconut palm are used also for awnings, deckhouses, weather screens and occasionally as bailers. The persistent leaf sheath of the coconut may also be sewn to form a cloth sail in the Sepik and Aitape areas of Papua New Guinea. Coconut palm fibre may be used for lashings and bindings but these are usually of rattan (*Calamus* spp.) or *Lygodium circinnatum,* a tough fern. Leaf midrib fibres of *Arenga saccharifera* are used in the Geelvink Bay area of Irian Jaya. The lianes *Calamus* sp. and *Flagellaria indica* make strong anchor ropes.

Long-lasting rafts are made from bamboo with rattan lashings, but any light buoyant wood may be employed and even banana stems (fastened with lianes) are used if the rafts are not required for more than one or two crossings.

House building

House styles and dimensions vary considerably from small round or rectangular family houses (Plate 49) and men's club-houses to community long-houses and highly decorated *haus tambaran* or other spirit-cult houses. Specially built and decorated yam storehouses are found in some areas also. Simple rain shelters, cooking shelters and outhouses are associated with homes and gardens in most parts of the country (Cranstone 1961, 1972; Meggitt 1957). Table 3.17 lists the species used in house construction, Table 3.18 some of the ropes used for lashing and binding of houses, other buildings and fences.

Coastal and lowland houses are often built above the ground, supported on piles 1-6 m high (Plate 49). The floors of split palm wood (*Archontophoenix* sp., *Nypa fruticans*, *Caryota rumphiana* among others) or *Pandanus* spp. are built first, and are supported on numerous short poles. Large centre-line posts and corner posts support the ridge-pole and wall plates respectively; rafters are tied to these and to intermediate wall posts where necessary. Tie-beams are added to help support large roof areas. Finer purlins are tied on to rafters and beams; these carry the roofing materials.

In highland areas houses are smaller and built on the ground or very close to it (Plate 50). They may be circular with a central post and radiating rafters tied on to wall planks or a wall beam, or rectangular with a ridge-pole and corner posts to support the rafters. In some areas the roof is hipped to provide extra space at the back end, which includes the sleeping quarters. The walls are built first, a double line of sharpened planks being driven into the ground and the gap between them filled with *Miscanthus floridulus*, other grasses and occasionally fern fronds, as insulation.

Relatively long-lasting hard and medium hardwoods are preferred for house posts and beams (*Calophyllum inophyllum*, *Dysoxylum* spp., *Elaeocarpus* spp., *Euodia* spp., *Casuarina* spp., *Castanopsis acuminatissima*, *Nothofagus* sp., among others, Table 3.17) while lighter woods (*Acalypha* spp., *Barringtonia racemosa*, *Antidesma* sp., *Dodonaea viscosa*, *Ficus* spp., etc.) form the rafters and purlins. Bamboo is used whenever readily available for house beams, intermediate wall posts, rafters and purlins. Slender-stemmed species are used for house blinds and thatching rods in New Britain and many other coastal and lowland areas. In the highlands the large-leaved forms of *Bambusa forbesii* may be used as roofing shingles (Floyd 1954; Clarke 1971).

In coastal areas house walls, where they exist, may be made of sago or nipa palm leaves, upright sago stems, sheets of bark, *Pandanus* leaves or wooden planks (Chowning, pers. comm.; Held 1957; Williams 1930, 1940). In highland areas wooden planks are used for walls and both inner and outer walls may be lined with bark of *Buchanania arborescens*, *Papuacedrus papuana*, Lauraceae or *Pandanus* spp., or with leaves of the latter for added warmth. House roofing is usually of sheets of atap, that is, dried leaves of sago or *Nypa* palm, stitched to bamboo laths of 1.5-2.0 m length, in many coastal and lowland areas, but may be made from *Calamus* sp. cane, *Pandanus* spp. leaves or *Dendrocnide excelsa* (*Laportea gigas*) bark.

Plate 49
Lowland village near Kiunga, Western District. The small size of the planted breadfruit, pawpaw, coconut and banana indicates recent establishment.

Plate 50
A typical highland house with plank walls and grass thatch

Table 3.17 Plants used in house, shelter and fence building and decoration

Plant name	Family	Use	Area recorded
Acalypha insulana	Euphorbiaceae	Housebuilding timber	Tari
Acalypha sp.	Euphorbiaceae	Flexible wall joints, general	Wissel Lakes, Chimbu
Aglaia sapindina	Meliaceae	House posts	New Britain
Alphitonia incana	Rhamnaceae	Housebuilding timber	Jimi, Tari, Chimbu
Alphitonia moluccana	Rhamnaceae	House beams	New Britain
Alpinia spp.	Zingiberaceae	Roofing	Jimi
Alstonia sp.	Apocynaceae	Beams	New Britain
Antidesma sp.	Euphorbiaceae	Posts	New Britain
Aphanamixis sp.	Meliaceae	Posts	New Britain
Archontophoenix sp.	Palmae	Flooring, roofing	New Britain
Ardisia sp.	Myrsinaceae	House building	Chimbu
Arthraxon ciliaris	Poaceae	Thatch	Chimbu
Ascarina philippinensis	Chloranthaceae	Housebuilding timber, leaves as decoration	Chimbu
Astronia sp.	Melastomataceae	Housebuilding timber	Chimbu
Bambusa forbesii	Poaceae	Roofing material — leaves	Highlands
Bambusa spp.	Poaceae	General housebuilding material	Widespread
Barringtonia aff. *racemosa*	Barringtoniaceae	Beams	New Britain
Boerlagiodendron sp.	Araliaceae	Building materials	Jimi
Breynia sp.	Euphorbiaceae	Housebuilding	Tari
Buchanania arborescens	Anacardiaceae	Bark for house walls	Jimi
Calamus spp.	Palmae	Leaves as house wall insulation	Hewa
Callicarpa pentandra	Verbenaceae	Beams	New Britain
Calophyllum inophyllum	Clusiaceae	Posts	New Britain
Carpodetus sp.	Saxifragaceae	Housebuilding timber	Chimbu
Caryota rumphiana	Palmae	Flooring	New Britain
Castanopsis acuminatissima	Fagaceae	Planks, posts, leaves as decoration, bark as insulation	Tari, Wissel Lakes, H Chimbu, Enga
Casuarina equisetifolia	Casuarinaceae	Beams	New Britain
Casuarina oligodon	Casuarinaceae	Posts, planks, ridge poles, rafters, beams, bark as insulation	Enga, Chimbu, Tari
Casuarina papuana	Casuarinaceae	Posts, beams, ridgepoles, rafters	Jimi, Tari
Celosia argentea	Amaranthaceae	House decoration	—
Chisocheton sp.	Meliaceae	House planking	New Britain
Cinnamomum sp.	Lauraceae	Timber	Chimbu
Claoxylon sp.	Euphorbiaceae	Building timber	Chimbu
Conandrium polyanthum	Myrsinaceae	Small rafters	Tari
Cordyline sp.	Liliaceae	Leaves knotted together as measuring rope	Enga
Costus sp.	Zingiberaceae	—	Jimi
Croton sp.	Euphorbiaceae	Building materials	Chimbu
Cryptocarya sp.	Lauraceae	Housebuilding timber	Chimbu
Daphniphyllum sp.	Daphniphyllaceae	Housebuilding timber	Chimbu
Decaspermum sp.	Myrtaceae	Building materials	Jimi, Chimbu
Dendrocnide excelsa (syn. *Laportea gigas*)	Urticaceae	Roofing of outhouses	New Britain
Dillenia sp.	Dilleniaceae	Building materials	Jimi
Dodonaea viscosa	Sapindaceae	Housebuilding timber	Tari, Jimi, Chimbu
Donax canniformis	Marantaceae	Bark used for sewing thatch	New Britain
Drymaria rigidula	Caryophyllaceae	Leaves	Chimbu
Dysoxylum spp.	Meliaceae	Posts	New Britain
Elaeocarpus spp.	Elaeocarpaceae	Beams	New Britain, highlan
Euodia bonwickii	Rutaceae	Beams	New Britain
Euodia crassiramis	Rutaceae	Beams	New Britain
Euodia sp.	Rutaceae	—	Chimbu
Eurya spp.	Theaceae	Housebuilding	Tari, Chimbu
Ficus adenosperma	Moraceae	Housebuilding	Chimbu
Ficus trichocerasa	Moraceae	Housebuilding timber	Chimbu

Plant name	Family	Use	Area recorded
Ficus spp.	Moraceae	Aerial roots as housebeams, general building timber	New Britain, Chimbu
Gahnia sieberiana	Cyperaceae	Leaves as thatch	Wissel Lakes
Garcinia sp.	Clusiaceae	Rafters	Tari
Gironniera sp.	Ulmaceae	Building materials	Jimi
Glochidion pomiferum	Euphorbiaceae	Housebuilding timber	Tari
Glochidion spp.	Euphorbiaceae	Housebuilding timber	Chimbu
Gordonia papuana	Theaceae	Housebuilding timber	Chimbu
Grevillea papuana	Proteaceae	Housebuilding timber, leaves	Chimbu
Helicia microphylla	Proteaceae	Leaves	Chimbu
Helicia oreadum	Proteaceae	Housebuilding timber	Chimbu
Heliconia sp.	Heliconiaceae	Decoration	Jimi
Hibiscus tiliaceus	Malvaceae	Posts	New Britain
Homalium foetidum	Flacourtiaceae	Posts	New Britain
Horsfieldia sp.	Myristicaceae	Beams	New Britain
Imperata cylindrica	Poaceae	Roofing of houses	New Britain, widespread highlands
Intsia bijuga	Leguminosae	Posts	New Britain
Ischaemum polystachyum	Poaceae	Thatch, garden shelters	Tari
Leea sp.	Vitaceae	Building materials	Jimi
Leucosyke sp.	Urticaceae	Beams Inner bark for twine, house lashing	New Britain
Levieria sp.	Monimiaceae	Housebuilding timber	Chimbu
Lithocarpus cf. *rufo-villosus*	Fagaceae	Timber as posts, planks, leaves as decoration	Tari, Chimbu
Lithocarpus spp.	Fagaceae	Planks	Wissel Lakes, Jimi
Macaranga aleuritoides	Euphorbiaceae	Beams	New Britain
Macaranga tanarius	Euphorbiaceae	Beams	New Britain
Macaranga spp.	Euphorbiaceae	House timber	Widespread
Meliosma sp.	Sabiaceae	Building timber	Chimbu
Metroxylon spp.	Palmae	House roofing, midrib as walls	New Britain, Orokaiva
Miscanthus floridulus	Poaceae	Tie beams, thatch weights, general	Tari, Chimbu
Mischocodon sp.	Sapindaceae	Building materials	Jimi
Musa spp.	Monimiaceae	Roofing material	Jimi
Myristica sp.	Myristicaceae	Planks	New Britain
Nauclea sp.	Rubiaceae	Posts	New Britain
Neonauclea sp.	Rubiaceae	Housebuilding timber	Chimbu
Nothofagus spp.	Fagaceae	Planks	Wissel Lakes, Tari, Hagen, Chimbu
Nypa fruticans	Palmae	House roofing Floor slats	New Britain, Orokaiva Waropen
Octamyrtus arfakensis	Myrtaceae	Posts	Wissel Lakes
Olearia spp.	Asteraceae	Housebuilding timber	Chimbu
Omalanthus novoguineensis	Euphorbiaceae	Housebuilding	Tari
Omalanthus populneus	Euphorbiaceae	Beams	New Britain
Oreocnide sp.	Urticaceae	Building materials	Jimi
Palmeria sp.	Monimiaceae	Building material	Chimbu
Pandanus spp.	Pandanaceae	Wall lining	Tari
Papuacedrus papuana	Cupressaceae	Housebuilding timber	Chimbu
Phragmites karka	Poaceae	House construction	Widespread highlands
Phrynium sp.	Marantaceae	Roofing	Jimi
Phyllanthus archboldianus	Euphorbiaceae	Building timber	Chimbu
Phyllanthus flaviflorus	Euphorbiaceae	Large and small rafters	Tari
Phyllanthus spp.	Euphorbiaceae	Large and small rafters	Tari, Jimi, Chimbu
Pipturus sp.	Urticaceae	Houseposts	Tari
Planchonella sp.	Sapotaceae	Building timber	Chimbu
Poa brassii	Poaceae	Leaves as thatch	Chimbu
Podocarpus amarus	Podocarpaceae	Timber in buildings	Chimbu

Plant name	Family	Use	Area recorded
Podocarpus vitiensis	Podocarpaceae	Planks	Wissel Lakes
Polyosma sp.	Saxifragaceae	Building timber	Chimbu
Pometia pinnata	Sapindaceae	Planks	New Britain
Premna integrifolia	Verbenaceae	Posts	New Britain
Pterocarpus indicus	Leguminosae	Beams	New Britain
Quintinia sp.	Saxifragaceae	Planks	Tari
Rapanea sp.	Myrsinaceae	Timber in buildings	Chimbu
Rhizophora spp.	Rhizophoraceae	Posts	Waropen
Rhus sp.	Anacardiaceae	Building timber	Chimbu
Saccharum spp.	Poaceae	Roofing, purlins, fencing	New Britain, highlands
Saurauia spp.	Saurauiaceae	Building materials, fencing	Jimi, Chimbu, Hagen
Schefflera sp.	Araliaceae	Building timber	Chimbu
Schizomeria spp.	Cunoniaceae	Timber in buildings	Chimbu
Sloanea sp.	Elaeocarpaceae	Building timber	Chimbu
Smithia sensitiva	Leguminosae	Bark as fire-plate above hearth	Tari
Sonneratia alba	Sonneratiaceae	Beams	New Britain
Spiraeopsis celebica	Cunoniaceae	Building timber	Chimbu
Spiraeopsis spp.	Cunoniaceae	Building materials	Jimi, Chimbu
Sterculia sp.	Sterculiaceae	Beams	New Britain
Streblus urophyllus	Moraceae	Building timber	Chimbu
Symplocos sp.	Symplocaceae	Building timber	Chimbu
Syzygium spp.	Myrtaceae	Posts and beams	New Britain, Tari, Chimbu, Jimi, Wissel Lakes
Ternstroemia spp.	Theaceae	Building materials	Jimi, Chimbu
Timonius avensis	Rubiaceae	Building materials	Chimbu
Trema sp.	Ulmaceae	Beams	New Britain
Trimenia papuana	Trimeniaceae	Rafters	Tari
Trimenia sp.	Trimeniaceae	Building timber	Chimbu
Vaccinium sp.	Ericaceae	Building timber	Chimbu
Vitex cofassus	Verbenaceae	Posts and planking	New Britain
Wendlandia paniculata	Rubiaceae	Posts, rafters and purlins	Tari, Chimbu
Xanthomyrtus sp.	Myrtaceae	Posts, beams, thatch purlins	Tari

Smaller outhouses and shelters have roofs thatched with *Saccharum* sp. or *Imperata cylindrica*. In the highlands houses are almost always thatched with bundles of kunai grass, *Imperata cylindrica*, although locally *Pandanus* spp. and *Papuacedrus papuana* barks are used. Smaller cookhouses, outhouses and shelters may be thatched with banana leaves, *Phrynium* sp., *Bambusa* spp., *Alpinia* spp., *Pandanus* spp. or *Miscanthus floridulus*, *Phragmites karka*, *Poa brassii* or *Arthraxon ciliaris*.

Ropes and vines are used in all construction jobs (Table 3.18). The latter are collected from grassland and forest areas and in some cases scraped and dried and coiled for a few weeks before building starts. Rattan is widely used for tying floors and for suspending the clay fire-hearth below the floor in above-ground houses.

House *bilas* (decoration) in highland areas comprises leaves of *Castanopsis acuminatissima* and *Lithocarpus* sp. fastened under the eaves and often animal jaw bones and other trophies. Ornamental plants are grown nearby: *Heliconia* sp., *Cordyline fruticosa*, *Codiaeum variegatum*, *Celosia argentea*, *Coleus scutellarioides* and *Plectranthus* sp. among others. Hedges of purple-flowered *Graptophyllum pictum*, red and green-leaved *Cordyline* sp., *Oreocnide* sp., *Saurauia* sp. and *Saccharum* spp. may be planted.

Most houses have little furniture. Raised sleeping platforms of bamboo and palm or other wood slats may be built, or two 150 cm long, 30 cm wide planks of wood supported by four short posts may serve as a bed. Head-rests are widely used; they may be simply a section of bamboo or be carved from tree branches and roots, sometimes they are highly decorated. Mats of coconut or sago palm leaf or bark cloth may be used as floor and bed covers in lowland areas while in the highlands bark cloth capes and *Pandanus* leaf rain-capes serve as covers. Fires are kept burning most of the night in high-altitude areas and here a bark plate over the hearth serves as a fire-baffle for the thatch and a drying rack for tobacco, crop seeds and fibres. Personal goods may be stored in hollowed-out logs of

Table 3.18 Plants used as ropes in construction of houses, shelters, fences, etc.

Plant name	Family	Use	Area recorded
Abroma augusta	Sterculiaceae	Lashing	New Britain
Adenia sp.	Passifloraceae	Lashing	Jimi
Aeschynanthus sp.	Gesneriaceae	Lashing	Jimi
Blumea riparia	Asteraceae	Rope used in light tying	Jimi
Calamus spp.	Palmae	Cane used for all types of tying, plaiting, weaving	Widespread
Cassytha filiformis	Lauraceae	Stem used for fastening roof	New Britain
Celastrus novoguineensis	Celastraceae	Lashing	Tari
Cyclosorus unitus	Thelypteridaceae	Lashing	Tari
Debregeasia sp.	Urticaceae	Lashing	Tari
Dicranopteris sp.	Gleicheniaceae	Lashing	Jimi
Dimorphanthera spp.	Ericaceae	Lashing	Jimi, Hagen, Chimbu
Elaeagnus sp.	Elaeagnaceae	Stem as rope	Chimbu
Embelia sp.	Myrsinaceae	Lashing	Tari
Flagellaria indica	Flagellariaceae	Rope for binding house roofing	New Britain, Jimi
Freycinetia sp.	Pandanaceae	Rope used in light tying	Jimi, Tari, Chimbu
Geitonoplesium cymosum	Liliaceae	Lashing	Tari, Chimbu
Gleichenia brassii	Gleicheniaceae	Rope for heavy tying, fence posts	Mt Hagen
Hoya sp.	Asclepiadaceae	Lashing	Tari, Chimbu
Imperata cylindrica	Poaceae	Used for light cordage	—
Ipomoea sp.	Convolvulaceae	General binding in house construction	New Britain
Lepistemon ureceolatum	Convolvulaceae	Vine used for light tying	—
Lucinaea sp.	Rubiaceae	Stem used as rope	Chimbu
Lygodium circinnatum	Schizaeaceae	Stem used for canoe lashings and baskets	New Britain, Bougainville, New Ireland
Medinilla sp.	Melastomataceae	Rope used for lashing houses, fences	Jimi, Tari
Metroxylon spp.	Palmae	Heavy stiff rope — general lashing	Waropen
Nastus spp.	Poaceae	Stem used for general tying purposes	Mt Hagen, Chimbu
Pandorea pandorana	Bignoniaceae	Stem used as rope	Chimbu
Parsonsia pedunculata	Apocynaceae	Rope used for fencing	New Britain
Pilea sp.	Urticaceae	Stem used as rope	Chimbu
Piper sp.	Piperaceae	Rope used to tie roof of houses	Tari
Pueraria spp.	Leguminosae	Stem as rope	Highlands
Rhamnus nepalensis	Rhamnaceae	Stem as rope	Chimbu
Rubus moluccanus	Rosaceae	Lashing	Jimi, Chimbu
Scaevola oppositifolia	Goodeniaceae	Stem as rope	Chimbu
Schizomeria sp.	Cunoniaceae	Stem as rope	Chimbu
Syzygium sp.	Myrtaceae	Lashing — bark	New Britain
Triumfetta pilosa	Tiliaceae	Lashing, pig rope — bark	Enga, Tari
Triumfetta rhomboidea	Tiliaceae	Lashing — bark	Chimbu
Uncaria sp.	Rubiaceae	Stem as rope	Chimbu
Vaccinium cf. *auriculifolium*	Ericaceae	Lashing	Jimi

Pandanus spp. lashed to the side walls of the house or in the thatch. Brooms are made from the inflorescences of coconut and other palms, a section of the leaf midrib or roots of the coconut palm or a whole plant of *Selaginella caudata* (Floyd 1954).

Food preparation, containers and vessels

Food may be roasted, baked, steamed or boiled or cooked between heated stones. Frequently the staple starchy tubers are roasted by the side of the fire, the supplementary green vegetables and sweetmeats, if available, steamed in a bamboo container (30-45 cm long, 2-3 cm in diameter) in leaves or sometimes a bark roll, inserted into the hot ashes. In some coastal and lowland areas where clay pots are available food may be boiled or steamed in these. Coconut shells are used occasionally. Above-ground ovens and earth ovens are also used. The earth oven consists of a hole, usually 1.0-1.5 m in diameter and 0.5-1.0 m deep, lined with leaves. Stones, heated on a nearby wood fire, are placed in the base of the oven and then food packages are laid on top. Further stones are added, water is sprinkled on top and the whole covered with leaves, occasionally bark, and soil, and left to steam. The hot stones are handled with tongs: a strip of bamboo cortex, doubled on itself when green (Williams 1936; Landtman 1933; Blackwood 1950), or a split wooden stick. Banana leaves are used most often for lining earth ovens and for wrapping the food to be cooked but many other leaves can be used (Table 3.19) and often replace banana leaves on ceremonial occasions. Some such as *Alpinia* spp. and *Barringtonia asiatica* impart flavour to the food, while others, including *Pipturus* sp., *Myristica* sp., *Ficus dammaropsis* and other *Ficus* spp. may be eaten. Bark of *Terminalia catappa, Pandanus* spp. and of Lauraceae may be used to cover the oven.

The *bilum* or string bag is the common container in all parts of mainland New Guinea; it is used for carrying food, personal effects, babies and piglets. In many coastal and lowland areas and in the islands plaited leaf baskets and woven bags are used; in the case of the former, the midrib of a sago or coconut palm leaf is split to form the basket margin and the leaflets interplanted below to form the basket. In some areas finely coiled baskets are made from coconut or *Pandanus* leaves, *Lygodium* fern fibre or stems of *Rubus moluccanus*. Vessels used for carrying food from forest or garden to the house include the sheathing leaf bases of Palmae (*Areca catechu, Archontophoenix* sp. and *Gronophyllum chaunostachys*) and broad leaves, such as those of the Araceae *Alocasia macrorrhiza* and *Cyrtosperma chamissonis*, and bananas. These are discarded after little use. Long sections (50-250 cm) of bamboo are used as water containers, shorter sections as containers for collecting insects, amphibians and small crustacea. The ends are stopped up with rolled leaves of sago or coconut palm, folded leaves of *Spathoglottis* sp. and other ground orchids and *Dianella ensifolia* and bunches of *Scleria* sp. or *Garcinia* sp. leaves among others. In the Kutubu area very long bamboos are used for storing and carrying the valuable tigasso oil, gathered from *Campnosperma* sp.

trees. A sago spathe, folded and sewn to form a water-tight vessel, and with a rattan handle, serves as a bucket in the Kutubu and Trans-Fly areas. In others taro leaves serve as water and honey containers (Chowning, pers. comm.), and dried coconut shells serve as water and lime storage vessels. A water dish for cassowaries may be carved from the stem of the tree fern *Cyathea angiensis*.

Coconut shells are used as food containers, bowls, and drinking cups, and banana leaves as plates. Wooden bowls are used quite widely and are often highly decorated with inlaid pearlshell and carved margins and handles. Gourds (*Lagenaria siceraria*) are grown in most areas for storage vessels. The ground varieties with broad flat bases are used for storage of pig fat, vegetable oil, bean seeds, salt and peanuts, while short narrow-necked varieties are kept for lime storage, associated with betel nut chewing. Long-fruited gourds are used mainly for water containers, occasionally for oil. In

Table 3.19 Leaves used for lining cooking ovens and wrapping food for cooking

Plant name	Family	Area recorded
Alocasia hollrungii	Araceae	Chimbu
Alocasia macrorrhiza	Araceae	Jimi
Alpinia spp.	Zingiberaceae	Widespread
Astronia sp.	Melastomataceae	Jimi
Barringtonia asiatica	Barringtoniaceae	New Britain
Calanthe sp.	Orchidaceae	Jimi
Coix lachryma-jobi	Poaceae	Chimbu
Cominsia sp.	Marantaceae	New Britain
Cordyline fruticosa	Liliaceae	Widespread
Crotalaria semperflorens	Leguminosae	Chimbu
Cyclosorus sp.	Thelypteridaceae	Chimbu
Dillenia sp.	Dilleniaceae	Jimi
Endospermum formicarum	Euphorbiaceae	New Britain
Fagraea racemosa	Loganiaceae	Jimi
Ficus dammaropsis	Moraceae	Highlands
Ficus spp.	Moraceae	Widespread
Gmelina moluccana	Verbenaceae	New Britain
Heliconia sp.	Heliconiaceae	Jimi, New Britain
Ischaemum polystachyum	Poaceae	Chimbu
Kleinhovia hospita	Sterculiaceae	New Britain
Lunasia amara	Rutaceae	Jimi, New Britain
Macaranga aleuritoides	Euphorbiaceae	New Britain
Macaranga tanarius	Euphorbiaceae	New Britain
Musa spp.	Musaceae	Widespread
Myristica sp.	Myristicaceae	Jimi
Omalanthus populneus	Euphorbiaceae	New Britain
Omalanthus spp.	Euphorbiaceae	Widespread
Paspalum conjugatum	Poaceae	Chimbu
Pennisetum macrostachyum	Poaceae	Chimbu
Phrynium sp.	Marantaceae	Jimi
Pipturus sp.	Urticaceae	Jimi
Pteris sp.	Pteridaceae	Chimbu
Riedelia carallina	Zingiberaceae	Chimbu
Saurauia spp.	Saurauiaceae	Chimbu
Schefflera sp.	Araliaceae	Tari
Spathoglottis sp.	Orchidaceae	Jimi
Syzygium sp.	Myrtaceae	Chimbu
Wendlandia paniculata	Rubiaceae	Chimbu
Xanthosoma sagittifolium	Araceae	Jimi

the Gulf District gourds are used exclusively for lime, according to Landtman (1933).

Cordage, barkcloth and other textiles

It is obvious from the preceding sections on material culture that cordage is an essential part of the New Guinean's tool-kit. String, produced from the inner fibre of many trees and shrubs, and bark beaten out into cloth, provide materials for clothing and ornamentation, for wrapping and tying personal possessions, for fish nets, bags and baskets (Table 3.20). The bark of *Hibiscus tiliaceus, Triumfetta pilosa* and *Syzygium* spp. may be simply stripped from the tree and twisted or plaited into ropes used for tethering pigs in garden areas or for rough tying purposes. The aerial root fibres of *Pandanus* spp. and the roots and husk fibres of the coconut palm are used in a similar manner.

Some species cultivated primarily for food (*Artocarpus altilis, Gnetum gnemon*) provide bark and fibre also, while others including *Broussonetia papyrifera, Debregeasia* spp. and *Phaleria macrocarpa* are planted especially for these products. Many Urticaceae are semi-domesticated; self-sown species of *Maoutia, Leucosyke, Pipturus, Pouzolzia* and *Boehmeria* are usually retained in gardens and clearings. *Ficus* spp. are used very widely and in some areas (Kukukuku, Kutubu) are the main source of fibre and bark. *Ficus iodotricha, F. pachyrrhachis* and *F. gymnorygma* are used for string while *F. trachypison, F. wassa* and *F. robusta* are suitable for bark cloth. The bark of many other species (Table 3.20) gathered in forest and regrowth areas, is taken back to the house for processing (Plate 51).

For making string the bark is carefully cut from the branch or trunk of the tree and the inner bark immediately peeled away from the outer layer. It is then dried in the sun for 1-2 days. With young saplings the outer periderm is scraped off with a bamboo knife and the stem then dried for a few days before stripping the fibre from the inside wood. The inner bark sheets are kept in the house for 1-2 weeks until completely dry and then shredded with the fingers or with bamboo tongs into short, thin fibres. To do this the large, hard strip is softened by soaking in water or by covering it with soft weeds such as *Crassocephalum crepidioides, Viola* spp. and *Amaranthus* spp. In the Kukukuku area the bark is chewed to soften it before drying (Blackwood 1950). String is produced by rolling two bundles of fibre together on the thigh or calf of the leg. Extra fibres are added until the desired length of string is obtained. Coarse string may be made by rolling large bundles of fibres together but this is difficult and more often finer strings are plaited together to produce the thicker ropes.

To make cloth, the inner bark is used while still wet, so that the fibres are loose and elastic. The bark sheet is laid over a tree trunk and beaten with a heavy wooden, or occasionally stone, club. Water is sprinkled on the cloth at intervals or a juicy banana petiole beaten over it. The material is folded upon itself and beaten again and again until a thin sheet of matted texture is formed. Designs cut into

Table 3.20 Plants used for making string and bark cloth

Plant name	Family	Use	Area recorded
Abroma augusta	Sterculiaceae	Bark fibre for string for clothing, net bags	New Britain
Agave sp.	Amaryllidaceae	Leaf fibre for string	Chimbu
Althoffia pleiostigma	Tiliaceae	Bark fibre for string	Chimbu
Artocarpus altilis	Moraceae	Bark cloth for clothing, rain capes etc.	New Britain
Boehmeria sp.	Urticaceae	Bark fibre for string for clothing, net bags etc.	Wissel Lakes, Mt Hagen
Broussonetia papyrifera	Urticaceae	Bark cloth	Widespread
Cocos nucifera	Palmae	Roots plaited into ropes, husk fibres twisted for fishing lines	Fly River, New Britain
Commersonia bartramia	Sterculiaceae	Bark fibre used for string for women's girdles, for headbands, for tying up bundles of food	New Britain, Chimbu
Cypholophus sp.	Urticaceae	—	Chimbu
Debregeasia sp.	Urticaceae	Bark fibre for string for women's garments, net bags, fishing nets, for pig ropes and tying houses	Wissel Lakes, Tari, Hagen, Chimbu
Desmodium sp.	Leguminosae	Bark fibre for string for clothing etc.	Wissel Lakes
Donax canniformis	Marantaceae	Bark used for sewing thatch	New Britain
Ficus caulocarpa	Moraceae	Bark fibre for string	Chimbu
Ficus dammaropsis	Moraceae	Bark cloth for men's head covering, fibre for string	Mt Hagen, Chimbu
Ficus glabella	Moraceae	Bark shredded for women's skirt	Kukukuku
Ficus gymorygma	Moraceae	Bark fibre for string for women's garments etc.	Wissel Lakes
Ficus iodotricha	Moraceae	Bark fibre for string	Jimi
Ficus microdictya	Moraceae	Bark fibre for string	Chimbu
Ficus pachyrachis	Moraceae	Sapling bark fibre for string	Jimi
Ficus robusta	Moraceae	Bark cloth	Jimi
Ficus trachypison	Moraceae	Bark cloth for men's head covering	Jimi
Ficus wassa	Moraceae	Bark cloth for men's head covering	Jimi, Mt Hagen
Ficus spp.	Moraceae	Bark fibre for string for fish nets, net bags etc., bark cloth	Widespread
Freycinetia sp.	Pandanaceae	Vine supplies fibre for ropes	Jimi
Gnetum gnemon	Gnetaceae	Bark fibre for string — cultivated	New Britain
Helicia odorata	Proteaceae	Bark fibre for string	Wissel Lakes
Hibiscus tiliaceus	Malvaceae	Bark fibre for string, woven into mats, baskets etc.	New Britain, Manus, Bosavi
Hibiscus sp.	Malvaceae	Bark used as rope to fasten pig in garden	Tari
Laportea sp.	Urticaceae	Bark fibre for string	Wissel Lakes
Leucosyke sp.	Urticaceae	Bark fibre for twine for house binding	New Britain
Maoutia sp.	Urticaceae	Bark fibre for string	Jimi, New Britain, Chimbu
Metroxylon spp.	Palmae	Shredded leaf fibre for string	Gulf D.
Musa spp.	Musaceae	Stem fibre for string, shredded for women's skirt, ropes for pigs	Fly River, Rai Coast, Tari
Pandanus spp.	Pandanaceae	Aerial root fibre for string	Waropen, Orokaiva, Tari
Phaleria macrocarpa	Thymeleaceae	Bark fibre for string for net bags — cultivated	Jimi
Pipturus spp.	Urticaceae	Bark fibre for string	Hagen, Tari, Chimbu
Pipturus cf. *verticillatus*	Urticaceae	Bark fibre for string for women's garments, fishing nets, string bags	Wissel Lakes
Pisonia sp.	Nyctaginaceae	Bark fibre for string	Wissel Lakes
Pouzolzia hirta	Urticaceae	Bark fibre for string	Wissel Lakes, Hagen, Tari
Sterculia sp.	Sterculiaceae	Bark fibre for string	Wissel Lakes
Streblus sp.	Moraceae	Bark fibre for string	Wissel Lakes
Syzygium sp.	Myrtaceae	Bark used for rope	New Britain
Trema spp.	Ulmaceae	Bark fibre for string	New Britain, Hagen, Chimbu
Triumfetta pilosa	Tiliaceae	Bark used as pig rope in garden	Tari
Urena lobata	Malvaceae	Bark fibre for string	Wissel Lakes
Wikstroemia androsaemifolia	Thymeliaceae	Bark fibre for string for net bags	Tari

Plate 51
Pandanus sp. fibre and ropes being sold at Tari market. *Pandanus* leaf raincape in centre, string bag of *Debregeasia* sp. and *Broussonetia papyrifera* on right. Crops being sold include *Setaria palmifolia* (left), parcels of sago (in foreground and on right), sweet potatoes and ginger (centre), peanuts in the background

the beater are impressed upon the cloth. The finished cloth may be dyed or dried and painted with vegetable products or clays (see section on Art). Bark cloth takes the place of textiles in many areas. Plaited mats of sago and coconut palm leaves are made in coastal and lowland areas, and *Pandanus* spp. leaves are often dried and plaited or sewn into mats and capes.

Clothing and personal ornamentation

While complete nudity is recorded for some areas, most people wear simple clothing. In many coastal and lowland areas the male dress comprises a perineal band and apron of bark cloth held by a girdle; in others leaves are worn. A tail-piece of frayed sago or coconut palm leaf may be worn also. In Eastern and Central Highland areas aprons are of bark cloth or of netted string or fibre cords arranged over a waist band or bark belt; *Cordyline* sp. leaves are worn at the back (Plate 52). Further west, bamboo, coconut or gourd peniscases are worn attached to a waist cord. In many areas women wear belts of string or bark and string aprons (Plate 53) together with leaves of various species (Panoff 1970), while in others, a skirt of shredded banana, coconut or sago leaves or banana stem fibres is worn. In the highlands the sedge *Eleocharis* sp. is often cultivated for women's skirts.

During periods of mourning women wear special clothing. Banana leaves and the leaves of a wild *Cordyline* are used in parts of New Britain (Panoff 1970, 1972), while a bark cloth cowl decorated with *Coix lachryma-jobi* seeds and a knitted string skirt completely covered with these seeds are worn by widows in the Orokaiva area (Williams 1930).

Waistbands of one or both sexes are often embroidered with black banana skin, *Thespesia peekeli* fibre, *Dicranopteris linearis* fibre, *Calamus* sp. cane or coloured orchid stems, including red *Bulbophyllum* sp. and yellow *Liparis* sp. Plaited fibre arm and leg bands are commonly worn and may be decorated in the same manner. Other plaited and woven garments, often highly decorative, are

Plate 52
Netted string aprons made from *Broussonetia papyrifera*, *Debregeasia* sp. or *Maoutia* sp. fibre worn by highland men. The bark belt is of *Prunus* sp.

Plate 53
Women's clothes in the highlands: a string skirt, string bag and bark cloth.

worn at ceremonies and on ritual occasions; these are described in most ethnographic studies and some of the weaving techniques used are analysed in K. Bühler (1948).

Leaves and flowers are used everywhere in everyday and ceremonial dress. Table 3.21 lists some of those recorded in the literature. Variegated-leaved species, and highly coloured, shiny or

Table 3.21 Leaves and flowers used in everyday and ceremonial dress

Plant name	Family	Area recorded
Alphitonia incana	Rhamnaceae	Widespread
Ascarina philippinensis	Chloranthaceae	Chimbu
Astronia sp.	Melastomataceae	Tari
Callicarpa caudata	Verbenaceae	New Britain
Calycacanthus magnusianus	Acanthaceae	New Britain
*Celosia argentea	Amaranthaceae	New Britain, Chimbu
Cinnamomum sp.	Lauraceae	Chimbu
Clerodendron paniculatum	Verbenaceae	New Britain
Codiaeum variegatum	Euphorbiaceae	Widespread
Coleus atropurpureus	Lamiaceae	Widespread
Coleus blumea	Lamiaceae	New Britain
†*Coleus* spp.	Lamiaceae	Widespread
Cordyline fruticosa	Liliaceae	Widespread
Cordyline sp.	Liliaceae	Widespread
Crinum macrantherum	Amaryllidaceae	New Britain
Croton sp.	Euphorbiaceae	Widespread
Cryptocarya spp.	Lauraceae	Highlands, widespread
†*Cymbopogon citratus*	Poaceae	Widespread
Cyperus pedunculosus	Cyperaceae	Chimbu
Decaisnina hollrungii	Loranthaceae	Chimbu
*Dendrobium spp.	Orchidaceae	Chimbu
Dracaena sp.	Liliaceae	New Britain
Drimys spp.	Winteraceae	Jimi, Mt Hagen
Drymaria rigidula	Caryophyllaceae	Chimbu
Elatostema sp.	Urticaceae	Chimbu
Elmerrillia papuana	Magnoliaceae	New Britain, Jimi
†*Euodia anisodora*	Rutaceae	New Britain
Euodia elleryana	Rutaceae	New Britain
†*Euodia hortensis*	Rutaceae	Kukukuku
Euphorbia pulcherrima	Euphorbiaceae	Chimbu
Freycinetia sp.	Pandanaceae	Tari
Garcinia teysmanniana	Clusiaceae	Kukukuku
Garcinia spp.	Clusiaceae	Tari
*Giulianettia sp.	Orchidaceae	Chimbu
Gleichenia hirta	Gleicheniaceae	Kukukuku, Mt Hagen
Homalomena sp.	Araceae	Jimi
Horsfieldia spicata	Myristicaceae	Tari
*Impatiens spp.	Balsaminaceae	Tari
Justicia sp.	Acanthaceae	New Britain
†*Kaempferia galanga*	Zingiberaceae	Widespread
Lipocarpha chinensis	Cyperaceae	Chimbu
Litsea spp.	Lauraceae	Mt Hagen, Jimi
Lycopodium spp.	Lycopodiaceae	Tari
Macaranga sp.	Euphorbiaceae	Chimbu
Mallotus paniculatus	Euphorbiaceae	New Britain
Maoutia sp.	Urticaceae	New Britain
Messerschmidia argentea	Boraginaceae	New Britain
Myristica longipes	Myristicaceae	Tari
†*Ocimum basilicum*	Lamiaceae	Widespread
Omalanthus sp.	Euphorbiaceae	Widespread
Papuacedrus papuana	Cupressaceae	Widespread highlands

Plant name	Family	Area recorded
Pennisetum macrostachyum	Poaceae	Mt Hagen, Chimbu
Phreatia sp.	Orchidaceae	Chimbu
Piper spp.	Piperaceae	Widespread
**Pipturus* sp.	Urticaceae	Tari
†*Plectranthus* sp.	Lamiaceae	Widespread
Premna sp.	Verbenaceae	New Britain
Rapanea sp.	Myrsinaceae	Tari
**Rhododendron* spp.	Ericaceae	Chimbu, Hagen
Riedelia monticola	Zingiberaceae	Tari
**Riedelia* sp.	Zingiberaceae	Chimbu
Saurauia spp.	Saurauiaceae	Chimbu, Mt Hagen
Schefflera spp.	Araliaceae	Mt Hagen, Tari, Chimbu
Vernonia sp.	Asteraceae	New Britain
Zingiber spp.	Zingiberaceae	Widespread

* flowers used (otherwise leaves)
† aromatic

Table 3.22 Other species used in personal adornment

Plant name	Family	Part used and purpose	Area recorded
Adenanthera pavonina	Leguminosae	Seed	—
Alpinia spp.	Zingiberaceae	Leaf fibres as skirt	New Britain
Areca catechu	Palmae	Dyed lining of spathe as hair tassel	Orokaiva
Arthropteris obliterata	Oleandraceae	Leaves around neck	Kukukuku
Bambusa spp.	Poaceae	Tubes hold hair, disks in nasal septum, ear lobes, combs	Waropen, Trans-Fly, Star Mts, New Britain
Calamus spp.	Palmae	Finely woven cane belt, plaited belt, armlets, headband of vertical strips, necklace, legbands	Widespread
Cardiospermum halicacabum	Sapindaceae	Round black seeds as necklace	Wissel Lakes, Kukukuku
Cocos nucifera	Palmae	Leaf strips frayed, plaited into hair as tassels	Trans-Fly
		Midrib as base for feather headband	
		Armlets of coconut shell, rolled leaf in nasal septum	Orokaiva
Coix lachryma-jobi	Poaceae	Seeds as necklaces, nose ornaments, headband and cloak decoration, baldrics	Widespread
Cynoglossum sp.	Boraginaceae	Seeds	Chimbu
Dendrobium spp.	Orchidaceae	Yellow stem in waistband, bark belt, women's neck rope embroidery, headbands	Trans-Fly, Jimi, Chimbu
Dicranopteris linearis	Gleicheniaceae	Black core of stem for armbands, belts	Jimi
Dimorphanthera sp.	Ericaceae	Root fibre woven into arm bands	Tari
Diplocaulobium sp.	Orchidaceae	Stem skin woven into arm bands and belts, women's girdles	Kukukuku, New Britain
Erigeron sumatrensis	Asteraceae	Stem	Chimbu
Melothria sp.	Cucurbitaceae	Fruits with incised decoration as necklace worn on back	Jimi
Musa spp.	Musaceae	Black skin as waistband, plaited bark belt embroidery; black seeds as ear ornaments, necklaces	Trans-Fly, Orokaiva
Pandanus spp.	Pandanaceae	Strips of leaf plaited about hair tuft, rolled leaf in ear lobe, head ornament	Trans-Fly, Gulf D., Star Mts
Papuechites aambe	Apocynaceae	Fibre in armbands	Manus
Pithecellobium cf. *sapindoides*	Leguminosae	Seeds as necklaces	Jimi
Planchonella sp.	Sapotaceae	Brown seeds as necklaces	Jimi
Pullea glabra	Cunoniaceae	Wood	Chimbu
Scirpus mucronatus	Cyperaceae	Stems for women's skirts	Tari
Ternstroemia sp.	Theaceae	Twigs	Chimbu
Vaccinium sp.	Ericaceae	Wood	Chimbu

silvery-backed leaves are sought for headdresses and for skirt or back-side *bilas* for dancing, and aromatic leaves are often worn in arm and leg bands on these occasions. Other personal ornaments (Table 3.22) include seed necklaces of *Cardiospermum halicacabum, Coix lachryma-jobi, Pithecellobium* cf. *sapindoides, Planchonella* sp. and *Musa* spp., decorated fruit necklaces of *Melothria* sp., ear and nose ornaments of bamboo or coconut shell, armlets of coconut shell and *Calamus* sp. Ferns, leaves and flowers are frequently worn in the hair. Hair tassels made from dyed spathe linings of *Areca catehu* are recorded from the Orokaiva area, while in many other regions strips of *Pandanus* spp. leaves are plaited into the hair. In the Star Mountains a complex, heavy, club-shaped head ornament is made from strips of *Pandanus* sp. leaf, wood and clayey red earth (Clarke 1971; Williams 1930, 1936; Floyd 1954; Pospisil 1963; Blackwood 1940; A. and A.M. Strathern 1971; Kooijman 1962).

A number of vegetable oils are used as hair and body dressings, including coconut oil, *Campnosperma brevipetiolata* (tigasso oil), *Aleurites* sp. (candlenut oil) and *Pandanus conoideus* oil; these are highly valued and often traded over great distances for use on ceremonial occasions.

Discussion

Altogether 1035 plant species representing 470 genera and 146 families are recorded above as used for a variety of purposes in New Guinea. Straatmans (1967) compared, in various categories, the numbers of plants used in New Guinea with those of Indonesia. Considering the inadequacies of the data there is little value in carrying such comparisons further; the figures for New Guinea presented here alter considerably those given by Straatmans, for example 332 species (215 genera, ninety-nine families) of medicinal plants are recorded here as against eighty-eight by Straatmans, 251 species of food plant (in 150 genera, eighty-one families) as against 185, and a further sixty species recorded as fibre and rope plants. With increased interest in useful plants the present figures will be altered further in the future. Of the 146 families noted, sixty-three have only a single genus represented and a further thirty-five have only two genera recorded as useful. Well represented families include the Leguminosae, with twenty-four genera, important as food and also as medicines; the Urticaceae (sixteen genera) valued for their fibre and bark and also as subsidiary foods; and the Asteraceae (seventeen genera) used mainly for medicines. Euphorbiaceae (nineteen genera) are important as medicines and as soft woods for constructional purposes and handicrafts. The Palmae (twelve genera), Poaceae (twenty genera), and Araceae (thirteen genera) provide some of the staple food plants and also many of the raw materials needed for the New Guinea life-style. Of the families less well represented the Moraceae provide food, fibre and bark, latex and useful timber, the Convolvulaceae a staple food, the sweet potato, and vines and rope, the Cucurbitaceae, Commelinaceae and Acanthaceae supplementary foods. The Anacardiaceae and

Rutaceae provide some food and resins or latex, the Malvaceae and Sterculiaceae fibres and bark, and, together with the Meliaceae, Myrtaceae and Fagaceae, valuable timber for technological purposes. The Liliaceae, Melastomataceae, Orchidaceae and Lauraceae are ornamental and ritual plants, the Zingiberaceae mainly medicinal and ritual plants.

Individual genera of great value include *Dioscorea,* providing the edible yams, *Musa* and *Pandanus* providing food and many raw materials, *Canarium, Burckella* and *Terminalia,* as fruit and nut trees, *Gnetum* and *Bruguiera* as food, and fibre and timber plants respectively. Many of the genera and species listed are used for more than one purpose. Thus 249 of the 470 genera recorded, or 297 of the 1035 species, are multipurpose. Some of these plants provide man with basic sustenance and all material needs; in New Guinea genera such as *Cocos, Musa, Pandanus, Metroxylon* and possibly *Bambusa* fall into this category.

Subsistence agriculture: the plant base

Some 251 species of food plant are listed in Table 3.1; of these forty-three (17 per cent) are always cultivated, a further fifty-one (20 per cent) both cultivated and harvested as a wild resource, and 157 (63 per cent) gathered from the forests, savannas and grasslands. Lea (1975) considering the role of the tropical forest in providing sustenance points out the 'small but significant contribution' the gathering of wild plants and small fauna makes to most New Guinea societies. Although a few predominantly hunting and gathering groups are known in New Guinea today (Murdock 1967; Brookfield with Hart 1971) the majority of the people are subsistence agriculturalists growing mainly tuber crops and planting some fruit and nut trees. Considering the cultivated plants (Table 3.1), one is struck by the fact that most of the staples are introduced from elsewhere. Thus the yams (*Dioscorea* spp.) grown in New Guinea are possibly all South-east Asian in origin, the taro (*Colocasia esculenta*) of Indian or Indonesian origin and the sweet potato (*Ipomoea batatas*) of South American origin. The exceptions are sago (*Metroxylon* sp.) and some of the bananas (Australimusa section); other bananas (Eumusa section) are introduced from South-east Asia. Among the supplementary food crops some, such as the beans *Psophocarpus tetragonolobus* and *Dolichos lablab,* have been introduced from Asia, *Phaseolus lunatus* from America and the gourd, *Lagenaria siceraria,* from Africa (Whitaker 1971). Many more, however, are indigenous; examples are sugarcane (*Saccharum officinarum*), the edible grasses *Saccharum edule* and *Setaria palmifolia* and many green vegetables including *Rungia klossii, Hibiscus manihot, Oenanthe javanica, Commelina* spp., *Solanum nigrum,* and possibly the *Amaranthus* spp. Little is known of the origin of either the coconut (*Cocos nucifera*) or breadfruit (*Artocarpus altilis*) but some authorities consider the Melanesian area, from New Guinea to Fiji, their area of domestication (Purseglove 1972). Of the New Guinean plants utilised some are gathered from the wild state, others are tended in the forest or transplanted from the forest or grassland and cultivated, and others again are true domesticates, reproduced purposefully by man by vegetative means or by seed.

Archaeological evidence indicates that man has been in the Highlands of Papua New Guinea for at least 11,000 years and in areas marginal to the Highlands for 26,000 years (Bulmer 1966; White et al. 1970; White 1972); he probably reached the coastal areas 50,000 years ago or earlier. When man first arrived in New Guinea then, he must have been a hunter and gatherer, only more recently becoming an agriculturalist. This immediately poses such questions as what resources were available to the early immigrant, how and when agriculture began in New Guinea and along what lines it developed. Did it begin in New Guinea independently of the South-east Asian development or follow on the early introduction of the South-east Asian cultigens such as yams, taro and gourds? At present we do not have the evidence from either New Guinea or South-east Asia to answer these questions. Recent studies on the flora (Part I) and on the vegetation of New Guinea and its history (Flenley 1967; Powell 1970b; Walker 1970, 1972, 1973; Hope 1973) and reconstructions of former climates (and vegetation) in the Torres Strait area (Nix and Kalma 1972) and high-altitude areas (Hope and Peterson 1974), together with the more direct archaeological evidence of agriculture in the Highlands (Golson et al. 1967; Powell 1970a; Golson 1974a; Powell et al. 1975) and a linguistic study (Chowning 1963), provide a basis for speculation.

Former vegetation and environment: a hypothetical reconstruction

The natural flora of New Guinea today comprises widespread tropical, Malesian and Australian elements. On the basis of geological evidence it is considered that most of the extant families of the mountain areas (of southern affinity) would have reached New Guinea by the end of the Tertiary, and those of the lowlands somewhat later, in the Quaternary (Walker 1972). If this is so then diversification and speciation must have occurred rapidly during the Quaternary to explain the present endemism at both genus and species levels. Palynological evidence is lacking at present for these earlier periods.

Well before the time of man's entry, however, the vegetation was probably differentiated into distinct types, some of which may have been similar to those described by Paijmans in Part II. At the postulated time of man's entry into New Guinea, probably from Malesia, lower sea levels resulted in the development of land bridges between many of the Malayan and Indonesian islands and between New Guinea and Australia. Sea crossings, although still required in order to reach New Guinea, would have been shorter and possibly less hazardous than before. Nix and Kalma (1972) consider that the climate was cooler and somewhat drier. According to them the broad plain south of the central cordillera was probably covered with open forest and savanna, seasonally inundated woodland and fresh water swamps, while inland tropical rain forest clothed the foothill and mountain slopes up to about 300 m altitude and more temperate forests were present above this. By 17,000-14,000 B.P., the period of lowest sea levels, there was a further increase in aridity, and open sclerophyllous forest may have been present in the Markham, Ramu and lower Sepik valleys (Nix and Kalma 1972). During the whole of this period and up until circa 10,500 years B.P. the tree line was probably situated between 2000-2400 m altitude; shrub-rich grasslands occurred above this up to 3000-3200 m altitude and were replaced there by 'alpine' grassland; glaciers were extensive on the high peaks (Hope and Peterson 1974; J. and G. Hope 1974).

Thus it can be suggested (although direct evidence is not available), that the coastal strand vegetation and the lowland woodland and savanna country would have been extensive at the time of man's arrival and may have included some of the taxa still used there today: mangrove species, coconuts and other palms, bananas, breadfruit and *Pandanus, Terminalia, Barringtonia, Syzygium* spp. and possibly *Canarium,* sugarcane (*Saccharum officinarum*) and a variety of edible leaves including *Ficus* spp. and possibly *Gnetum gnemon* among others. Areas of sago may have been present in low-lying areas of the Arafura shelf and along swampy river mouths. Coming from Malesia some of these taxa would be familiar to the early immigrants; the Palmae and some types of bananas and Pandanaceae were known (Chowning 1963). *Cycas* spp. seeds and pith and *Cordyline* spp. tubers may also have been important plant foods in ancient times (Barrau 1965). These taxa alone would have been enough to sustain life, and if one adds to them a fauna of marsupials and other mammals in the lowland woodlands and savannas, and fish, shellfish and crustacea present in riverine and coastal waters, then the immigrants were fairly well provisioned. Bulmer (1968) records the distribution of animal resources today in New Guinea and the strategies of hunting. These include the use of drives, with or without fire, which are important in dry woodland and savanna areas today; they may have been more important formerly. If larger marsupials were present, as suggested for higher-altitude areas by J. and G. Hope (1974) and known for regions of Australia at that time, the potential for hunting would have been much more favourable than at present.

The tropical rain forest may have been more restricted in distribution, but still offering many edible and otherwise useful plant species and at least some animals. The montane forests were possibly rich in fruit and nut trees including *Pandanus, Elaeocarpus, Castanopsis* and *Sterculia.* The tuberous *Pueraria lobata,* edible ferns and edible leaves of woody species (*Ficus* spp., Urticaceae in particular) and herbs (*Setaria palmifolia, Rungia klossii, Oenanthe javanica* amongst others) may have been available also.

J. and G. Hope (1974) have emphasised the value of the shrub-rich grasslands lying altitudinally immediately above the montane forests at this time and the forest-grassland ecotone communities. They consider that, in many ways, it was easier for people to move through and forage in such areas than to establish themselves initially in the closed forest zones. Probably people hunted and gathered over a wide range of ecological zones, depending upon access routes and availability of seasonal plant resources and game. A similar life-style is shown by some of the upper Strickland and Fly River people today.

After circa 14,000 B.P. the climate ameliorated, the glaciers melted and sea levels rose. Nix and Kalma (1972) suggest that by 8000 B.P., just before Torres Strait was flooded for the last time, the climate was warmer and wetter; swampland and tropical rain forest may have expanded over areas formerly occupied by open forest and savanna. At higher altitudes the tree line moved upwards, reducing the area of shrub-rich grassland available to fauna and to the hunters and gatherers. The Hopes (1974) suggest that by 9000 years ago the large marsupials were extinct and the people more dependent upon plant resources. Although evidence is lacking, it seems likely that well before this, people were tending their swampland and forest plants, clearing around the bases of breadfruit and other fruit and nut trees, thinning out and perhaps transplanting sago palms. Discarded banana stems, apical shoots of sugarcane, stems of *Hibiscus manihot,* breadfruit and fibre plants (Urticaceae, Moraceae) may have established themselves in debris near house shelters. Whether or not some form of subsistence agriculture based on local, semi-domesticated and domesticated species, was being practised by the time the South-east Asian cultigens were transferred to New Guinea is unknown. When did these cultigens arrive?

Taro, of Indian or Indonesian origin, was taken in early times to China and thence to Japan (Purseglove 1972) and also from South-east Asia into the Pacific. Chromosome numbers suggest there has been a two-stage transfer from South-east Asia to the Philippines, Melanesia and Polynesia, the diploid earlier form being widely distributed, the triploid form more restrictedly (Yen and Wheeler 1968). Just when taro reached New Guinea remains unknown but it must have been one of the earliest crops introduced. Almost certainly it entered before 4500 B.P., the date set for the expansion of rice out of South China, and it may have been much earlier (Yen 1971b). Cytological work on the Eumusa series of bananas suggests that they were transferred to New Guinea at a very early date also (Simmonds 1959). Some of the yams may in fact be indigenous to New Guinea (*Dioscorea bulbifera, D. nummularia, D. pentaphylla, D. hispida*) and, if so, would have been very important in the dry savanna areas of the southern plain as an early food source; alternatively, they could have entered early from South-east Asia or Indonesia; others (*D. alata, D. esculenta*) came in somewhat later (Burkill 1960; Purseglove 1972).

The other staple tuber crop in New Guinea today, the sweet potato, is of American origin. There is archaeological evidence for it at 4500 B.P. on the central Peruvian coast and at 8000-10,000 B.P. in the Chilca area south of Lima. It was known in Polynesia in pre-historic times, possibly by the second to eighth centuries A.D., and may have reached the eastern Melanesian islands from there (Yen 1971a). In Europe the plant was unknown until Columbus took it to Spain in 1492. Plant variation studies indicate three separate lines of introduction from America to Polynesia, Melanesia and Asia (Yen 1971a) and this is supported, in part, by the historical records.

Thus the Spaniards took it to Guam and the Philippines from Mexico in the sixteenth century, while the Portuguese took it from Brazil to their settlements in Africa and Asia (including India and the East Indies) in the fifteenth and sixteenth centuries. Probably from there it reached western Melanesia. If the sweet potato was in the eastern Melanesian islands much earlier, however, it may have reached New Guinea from that direction. Considering the amount of variation in the New Guinea varieties and the fact that seed production occurs frequently (seedlings are being continuously added to the highlanders' suite), the few originally introduced varieties may well have been swamped by later introductions and by outbreeding.

The establishment of taro, yam and bananas in rain forest communities, which were possibly widespread by 8000 years B.P. (Nix and Kalma 1972), would have been relatively simple especially if some form of subsistence agriculture was already being practised. If an earlier date, perhaps 9000-11,000 years B.P. (the earliest date for South-east Asian plant-man associations so far; Gorman 1969; Yen 1971b), is assumed for their transfer, then the yams and bananas, more tolerant of arid conditions than taro, probably spread more rapidly than it.

Between 8000-5000 B.P. the climate was warmer than at present and the forest tree line lay above its present position (J. and G. Hope 1974). Assuming it also remained wet, such a climate would favour the expansion of the tropical tuber crops, bananas and sugarcane into the highland areas.

Archaeological and palynological evidence for agriculture

The first archaeological evidence of agriculture in the Highlands does, in fact, date to the period 8000-5000 B.P. Thus, at Kuk, an archaeological site in the upper Wahgi valley, near Mt Hagen, three systems of drainage ditches are recorded at different levels in the swamp sediments. Radiocarbon dates indicate that the systems were in operation sometime before 6000 years B.P., between 2000-2400 years ago and 1200 years B.P., and from 300 years ago up until the 1920s. The shape and size of the ditches and their distribution pattern differ in each phase; by analogy with present-day systems found in the Wahgi valley and elsewhere, it is possible to interpret these differences in terms of the growing of various crops with specific cultivation techniques in a swampland environment. Thus it has been suggested that during the earliest period of cultivation taro was grown within the ditches themselves and other mixed vegetables, bananas and sugarcane were planted in the intervening drained or partially drained beds. The second phase of gardening, between 2400 and 1200 years ago, was more intensive; taro was planted on the plots and, while there may have been some interplanting of other crops, these were probably mainly grown in the drier slope gardens nearby. The latest gardening phase was adapted to the cultivation of sweet potato in raised beds on the swampland (Golson 1974a, b; Powell *et al.* 1975).

Fossil plant evidence associated with this site includes seeds, extracted from sediments from the levels of the ditches and immediately below them, suggesting that gardening was being practised on, or at least in the immediate vicintiy of, the swamp throughout the period with which we are concerned. Represented are plants such as *Solanum nigrum* and *Rubus rosifolius* commonly associated with gardens and garden fallows, and some, such as *Cyperus melanospermus, Ludwigia octavalvis* and *Paspalum conjugatum,* that are known from fallow on drained swampland. With the exception of sweet potato tubers, recovered from house site hearths associated with, and more recent than, the latest phase of drainage, none of the major crop plants presumably grown have been found.

Pollen analytical work at other sites in the Mt Hagen area (Manton, Draepi, Minjigina) records clearance of forest at least 5200 years ago (Powell 1970a, b). Initially degraded oak (*Castanopsis* spp.) and beech (*Nothofagus* spp.) forests were present on the valley slopes bordering the swamplands, rather than wide expanses of grassland. This suggests that some form of shifting cultivation was probably being practised on the hillslopes soon after the valley swamps were being utilised. Later, when gardening shifted back on to the swamps, the forest recovered somewhat on the slopes, but undoubtedly there was still some use of them as suggested above. Recorded from the agricultural horizon of the Manton site (dating from 2400 B.P. to 400-500 B.P.) are the domesticated taxa *Amaranthus* spp., *Rorippa* sp. and *Oenanthe javanica* and the semi-domesticated *Solanum nigrum, Commelina* sp., *Rubus rosifolius* and *Coleus* sp. As well many garden weeds are present (*Crassocephalum crepidioides, Ageratum* sp., *Drymaria cordata, Viola* sp., *Plantago* sp., *Haloragis* sp., *Polygonum* spp. and Lamiaceae) and tree and shrub species associated with gardens and settlement areas including *Casuarina* sp., *Dodonaea viscosa, Trema* sp., *Ficus wassa, Melastoma affine* and some Urticaceae.

Further pressure on the slopes dating from circa 1200 years ago is indicated in the pollen diagrams from the Minjigina site by the replacement of forest taxa with light-demanding trees and shrubs and grassland. This may be interpreted as registering the initial impact of the cultivation of sweet potato (Golson 1974b), a crop better suited to drier soil conditions than to swamps and tolerant of poorer soil conditions than taro. Later, a partial recovery of the forest is indicated; this is associated with the apparent conservation of some species (*Trema, Dodonaea,* Moraceae) and probable cultivation of others (*Casuarina,* Myrtaceae, Podocarpaceae). In fact, a more sophisticated pattern of land use appears to have developed, involving selective cutting of forests or secondary regrowth and conservation or planting of many different species (Powell 1970a, b; Golson 1974a; Powell *et al.* 1975).

From 5000 B.P. to the present the climate has deteriorated somewhat (J. and G. Hope 1974) and this has limited the upward extension of cultivation of the tropical tuber crops yams and taro. Sweet potato, being rather more tolerant of cool conditions than either taro or yams (although not resistant to frosts), has provided a staple crop that can be grown up to 2700 m altitude or more and has thus enabled settlement at these high elevations.

Flenley's (1967) evidence for forest reduction probably due to human activities at Lake Inim at 2530 m altitude at an inferred date of 1600 years ago may be associated with the cultivation of sweet potato there, and the Hopes' (1974) record of *Casuarina* being grown in Chimbu from circa 1200 years ago may be interpreted in similar terms. Such a sequence requires the sweet potato to arrive in New Guinea from the direction of Oceania rather than South-east Asia. While this is not generally accepted, there is no real evidence against it and such an entry is quite feasible if the early dates for its presence in Polynesia are correct.

Other useful plants and products

Many of the early food plants present in New Guinea (coconuts, sago, bananas, breadfruit, *Pandanus*) provided other useful products such as fibre, timber, thatch, mats, domestic utensils, medicines and personal clothing and ornamentation, and were probably used by the early immigrants. Of the narcotics, stimulants and intoxicants the palms, *Arenga* sp., *Areca* sp. and *Nypa fruticans* may have been present in lowland areas, although both *Arenga pinnata*, the sugar palm, and *Areca catechu*, the betel nut, are considered to be of Malesian origin by some authorities (Purseglove 1972). *Piper methysticum*, kava, may have been present also; if not, it was probably introduced from Polynesia, today the area of greatest production. Tobacco (*Nicotiana tabacum*), of American origin, probably reached New Guinea from the Philippines or Indonesia some time in the seventeenth century or possibly even earlier (Riesenfeld 1951).

At least some of the *Bambusa* spp. would have been present in New Guinea in early times but the large-stemmed varieties may have originated in Asia and been transferred from there by man (Burkill 1935). Many of the medicinal plants listed in Tables 3.5-13 are indigenous in New Guinea but also of widespread tropical distribution. The use of most of them, or of closely related species, as medicines in Asia, Malaya, the Philippines and Indonesia is documented in Burkill (1935) and suggests that medicinal plant knowledge has been built up over a long period of time. Almost all of the fibre plants listed in Table 3.20 are indigenous in New Guinea. One exception is *Broussonetia papyrifera*, of eastern Asian origin; this was cultivated in Java in early times, however, and was probably transferred to New Guinea from there.

Traditions of use of plants for ritual and magic, for decoration and for building, cooking and other technological purposes (Tables 3.14-19, 3.21, 3.22) may have been brought with the early immigrants but many more developed within the New Guinea environment. To extract sago efficiently from the palm and to prepare poisonous plants for eating requires an advanced level of technology. The techniques applied to maintain economic production of tropical crops at high altitudes are complex; neither these nor many other aspects have received the attention warranted to further our knowledge of plant utilisation and human adaptation.

Population: the nutritional base

Given the wide range of food plants available throughout New Guinea and the considerable time-depth of occupation, it is perhaps surprising that population densities are so low in many areas. Nutritional studies have indicated a wide range of dietary situations from those considered to be quite inadequate (in terms of both protein and calories) to those which are very satisfactory and well balanced (Hipsley and Clements 1950; Oomen and Malcolm 1958; Luyken and Luyken-Koning 1955; Couvée et al. 1962; Venkatachalan 1962; Bailey and Whiteman 1963; Hipsley and Kirk 1965). New Guinean average values for calorie and protein intake, 1880 calories per day and 10-30 g protein per day for highlanders and 1470 calories and 10-40 g protein per day for lowlanders, fall far short of those considered to be 'normal' elsewhere (Oomen 1971). However, the excellent physique and physical performance of many adults suggest that metabolic pathways may differ also from 'normal' and that New Guineans are well adapted to the food situation (Hipsley and Kirk 1965; Oomen and Corden 1970).

The diet of most people is monotonously starchy, with 70-90 per cent of the food intake consisting of one or a few of the staple crops (Oomen 1971). Few studies of crop production have been undertaken but potential yields of the tubers (yam, taro, sweet potato), the sago palm and bananas, are all 8-15 tons per ha (Oomen 1971; Barrau 1958; Lea 1966; Rappaport 1967; Clarke 1971). Sago on its own is an extremely poor food, comprising starch and perhaps 0.2 per cent crude protein, the latter often being lost during preparation procedures; large quantities must be eaten to satisfy calorie requirements. Taro, yams, sweet potato and bananas are widely considered to be nutritionally poor (Peters 1958; Hodges et al. 1950) but, according to Oomen (1971), they do contribute sufficient ascorbic acid, acceptable calorie levels and minimal levels of nitrogen and amino-acids (except for sulphur compounds) for adults, and their high moisture content increases otherwise low fluid intakes. For growing children and for pregnant and lactating women the protein intake would probably be inadequate. Supplementary crops (green vegetables, fruits and nuts) and animal resources must play a significant role in providing further protein, vitamins and minerals. While detailed production figures for the supplementary crops are almost entirely lacking, there is no doubt that their importance has been underrated in the past, at least for highland areas (cf. Rappaport 1967).

The distribution of these resources and of the staples depends to a large extent on the environmental conditions present. In some nutritionally marginal areas, such as those based on sago, the development of trading between coastal and inland groups, or between one part of the coast and another, has resulted in more adequate diets for both parties. Such trading involves social alliances, however, and these are easily upset. In other areas, such as Frederik-Hendrik Island, the development of complex gardening techniques ensures adequate food production. Equally complex techniques have made it possible to grow sweet potato at marginal altitudes in the highlands. In both cases production is easily upset by abnormal climatic conditions such as drought and frost.

In more favourable environments wild food resources may be abundant and vegetable and fruit production is relatively easy; with such a base population density can increase and excess food can be used for trading and ceremonial exchange. In the Maprik area and in the Trobriand Islands giant ceremonial yams up to 4 m long are grown for prestige and exchange (Lea 1966; Malinowski 1935) and in some parts of the highlands sweet potato production forms the base for complex pig exchange cycles (Rappaport 1967; Waddell 1972).

The present situation

At the time of European contact dense populations lived in the Chimbu, Enga and Maprik areas of mainland Papua New Guinea and possibly on the Gazelle Peninsula of New Britain. Elsewhere population densities were lower and there appeared to be little pressure on land for subsistence activities. Considering 1966 census data Ward (1970) suggests that in many areas the population density is well below that which the environment might support under a subsistence gardening economy. Conditions are changing rapidly today, however, in some parts of New Guinea. Populations are increasing at the rate of 2.8-3.0 per cent and considerable land pressure is being experienced partly as the result of changing subsistence patterns and the development of cash cropping based on rubber, cocoa, coffee and tea among other crops, but also because of former alienation of land for towns and plantations. While many new food crops (such as corn, cassava, peanuts, peas, french beans, tomatoes, lettuce, cabbage) have been introduced, some traditional foods, such as coconuts, have become more important as cash crops and others, including taro and yams, have been neglected. Natural forest resources are also under greater pressure than ever before, with large areas of forest being cleared; wild plant and animal foods and many other useful products are lost in the process and the local environments are altered considerably. Development problems and adaptive strategies are outside the scope of this contribution; they have been considered recently by a number of authors in Brookfield (1973) among others, but it must be emphasised that at least some of the present problems facing both rural and urban communities could be solved by the further development and exploitation, as both food and export crops, of plant resources already available within the country.

Bibliography

Anderson, J.A.R. 1964. Observations on climatic damage in peat swamp forests in Sarawak. *Commonw. For. Rev.*, **43** : 145-58.
Anell, B. 1960. *Hunting and trapping methods in Australia and Oceania.* Studia Ethnographica Upsaliensia 18, Almquist and Wiksell, Uppsala.
Anon. 1944. *Imperata cylindrica. Taxonomy, distribution, economic significance and control.* Imperial Agric. Bureaux Joint Publ. 7.
Anon. 1972. *Mangroves and man.* Aust. Cons. Found. Viewpoint no. 7, Parkville.
Archbold, R. and Rand, A.L. 1935. Results of the Archbold Expeditions No. 7 of 1935. Summary of the 1933-1934 Papuan Expedition. *Bull. Am. Mus. nat. Hist.*, **68**, Art. 8 : 527-79.
Ashton, P.S. 1965. Comments on 'Dry land forest formations and forest types in the Malay Peninsula'. *Malay. Forester*, **28** : 144-8.
Austen, L. 1945. Cultural changes in Kiriwina. *Oceania*, **16** : 15-60.
Austin, M.P., Ashton, P.S., and Greig-Smith, P. 1972. The application of quantitative methods to vegetation survey. III. A re-examination of rain forest data from Brunei. *J. Ecol.*, **60** : 305-24.
Bailey, K.V. and Whiteman, J. 1963. Dietary studies in the Chimbu (New Guinea Highlands). *Trop. Geogr. Med.*, **15** : 377-88.
Balgooy, M.M.J. van 1960. Preliminary plantgeographical analysis of the Pacific. *Blumea*, **10** : 385-430.
_____ 1969. A study on the diversity of island floras. *Blumea*, **17** : 139-78.
_____ 1971. Plant geography of the Pacific, as based on the distribution of Phanerogam genera. *Blumea*, Suppl. **6**.
_____ 1975. *Pacific Plant Areas*, Vol. III Rijksherbarium, Leiden.
Barbour, M.G. 1970. Is any angiosperm an obligate halophyte? *Am. Midl. Nat.*, **84**(1) : 105-20.
Barnes, A.C. 1964. *The sugar cane*. Hill, London.
Barrau, J. 1958. *Subsistence agriculture in Melanesia.* B.P. Bishop Museum Bull. 219, Bishop Museum Press, Honolulu.
_____ 1959. The sago palms and other food plants of marsh dwellers in the South Pacific Islands. *Econ. Bot.*, **13** : 151-62.
_____ 1965. Witnesses of the past: notes on some food plants of Oceania. *Ethnology*, **4** : 282-94.
Barth, F. 1971. Tribes and intertribal relations in the Fly headwaters. *Oceania*, **41** : 171-91.
Beard, J.S. 1967. An inland occurrence of mangrove. *West. Aust. Nat.*, **10**(5) : 112-15.
Beintema-Hietbrink, D. Floristic analysis of Ceylon. Unpublished.
Béziat, P. 1968. Quelques aspects de la végétation de la Cordillère de Merida (Andes Vénézuéliennes). *Bull. Soc. Hist. nat. Toulouse*, **104**(1/2) : 306-16.
Blackwood, B. 1939. Life on the Upper Watut, New Guinea. *Geogrl J.*, **94** : 11-28.
_____ 1940. Use of plants among the Kukukuku of Southeast Central New Guinea. *Proc. 6th. Pac. Sci. Congress*, **4** : 111-26.

_____ 1950. *The technology of modern stone age people in New Guinea.* Pitt Rivers Museum Occas. Papers on Technology 3, Oxford.

Blake, S.T. 1972. Idiospermum, *a new genus and family for* Calycanthus australiensis. Contr. Queensl. Herb. no. 12, Brisbane.

Bowers, N. 1964. A further note on a recently reported root crop from the New Guinea Highlands. *J. Polynesian Soc.,* **73** : 333-5.

Brass, L.J. 1938. Botanical results of the Archbold Expeditions XI. Notes on the vegetation of the Fly and Wassi Kussa Rivers, British New Guinea. *J. Arnold Arbor.,* **19** : 174-90.

_____ 1941a. Stone age agriculture in New Guinea. *Geogrl Rev.,* **31** : 555-69.

_____ 1941b. The 1938-39 expedition to the Snow Mountains, Netherlands New Guinea. *J. Arnold Arbor.,* **22** : 271-342.

_____ 1956. Results of the Archbold Expeditions No. 75 of 1953. Summary of the fourth Archbold Expedition to New Guinea (1953). *Bull. Am. Mus. nat. Hist.,* **111**(Art. 2) : 77-152.

_____ 1959. Results of the Archbold Expeditions No. 79. Summary of the fifth Archbold Expedition to New Guinea (1956-1957). *Bull. Am. Mus. nat. Hist.,* **118**(Art. 1) : 1-69.

_____ 1964. Results of the Archbold Expeditions No. 86. Summary of the sixth Archbold Expedition to New Guinea (1959). *Bull. Am. Mus. nat. Hist.,* **127**(Art. 4) : 149-215.

Brookfield, H.C. 1961. The highland peoples of New Guinea. *Geogrl J.,* **127** : 436-48.

_____ 1962. Local study and comparative method: an example from Central New Guinea. *Ann. Ass. Am. Geogr.,* **52** : 242-54.

_____ 1964. The ecology of highland settlement. Pp. 20-38 in J.B. Watson (ed.), *New Guinea: the Central Highlands. Am. Anthrop.,* **66**(4), Pt 2, Spec. Publ.

_____ (ed.) 1973. *The Pacific in transition: geographical perspectives on adaptation and change.* Aust. Nat. Univ. Press, Canberra.

_____ and Brown, P. 1963. *Struggle for land: agriculture and group territories among the Chimbu of the New Guinea Highlands.* Oxford University Press, London.

_____ with Hart, D. 1971. *Melanesia: a geographical interpretation of an island world.* Methuen, London.

Brown, M. and Powell, J.M. 1974. Frost and drought in the highlands of Papua New Guinea. *J. trop. Geogr.,* **38** : 1-6.

Bühler, A. 1948. Dyeing among primitive peoples. *CIBA Review,* **68** : 2477-512.

Bühler, K. 1948. Basic textile techniques. *CIBA Review,* **63** : 2290-320.

Bulmer, R.N.H. 1964. Edible seeds and prehistoric stone mortars in the Highlands of East New Guinea. *Man,* **64**(183) : 147-50.

_____ 1968. The strategies of hunting in New Guinea. *Oceania,* **38** : 302-18.

Bulmer, S. 1966. Pig bone from two archaeological sites in the New Guinea Highlands. *J. Polynesian Soc.,* **75** : 504-5.

_____ 1974. Settlement and economy in prehistoric Papua New Guinea. *Working Papers in Anthropology, Archaeology, Linguistics and Maori Studies* No. 30, Dept of Anthropology, University of Auckland.

_____ and Bulmer, R.N.H. 1964. The prehistory of the Australian New Guinea Highlands. Pp. 39-76, 162-82 in J.B. Watson (ed.), *New Guinea: the Central Highlands. Am. Anthrop.,* **66**(4), Pt. 2, Spec. Publ.

Burbidge, N.T. 1960. The phytogeography of the Australian region, *Aust. J. Bot.,* **8** : 75-211.

_____ 1963. *Dictionary of Australian plant genera: gymnosperms and angiosperms.* Sydney.

Burkill, I.H. 1935. *A Dictionary of the Economic Products of the Malay Peninsula*, 2 vols. Crown Agents for the Colonies, London.

_____ 1960. The organography and the evolution of Dioscoreaceae, the family of yams. *J. Linn. Soc. (Bot.)*, **56** : 319-412.

Burtt Davy, J. 1938. *The classification of tropical woody vegetation types.* Imp. For. Inst. Paper 13, Oxford Univ.

Carlquist, S. 1965. *Island life: a natural history of the islands of the world.* Garden City, New York.

_____ 1967. The biota of long distance dispersal. V. Plant dispersal to Pacific islands. *Bull. Torrey bot. Club*, **94** : 129-62.

Chippendale, G.M. 1972. Checklist of Northern Territory plants. *Proc. Linn. Soc. N.S.W.*, **96** : 207-67.

Chowning, A. 1963. Proto-Melanesian plant names. Pp. 39-44 in J. Barrau (ed.), *Plants and the migrations of Pacific peoples*. Bishop Museum Press, Honolulu.

Christian, C.S. and Stewart, G.A. 1953. General report on survey of Katherine-Darwin region, 1946. *CSIRO Aust. Land Res. Ser.*, **1**.

Clarke, W.C. 1971. *Place and people: an ecology of a New Guinean community*. Aust. Nat. Univ. Press, Canberra.

Conroy, W.L. and Bridgland, L.A. 1950. Native agriculture in Papua New Guinea. Pp. 72-91 in E.H. Hipsley and F.W. Clements (eds.), *Report of the New Guinea Nutrition Survey Expedition 1947*. Govt Printer, Sydney.

Coode, M.J.E. 1969a. *Manual of the forest trees of Papua and New Guinea. Part 1 Revised: Combretaceae*. Divn of Botany, Dept of Forests, Lae.

_____ 1969b. *A dictionary of the generic and family names of flowering plants for the New Guinea and southwest Pacific region*. Botany Bulletin 3, Dept of Forests, Papua New Guinea.

_____ and Stevens, P.F. 1972. Notes on the flora of two Papuan mountains. *Papua New Guin. Sci. Soc. Proc.*, **23** : 18-25.

Cooper, D. 1971. Some botanical and photochemical observations in Netherlands New Guinea, 'New Zealand New Guinea Expedition 1961'. *Econ. Bot.*, **25** : 345-56.

Corner, E.J.H. 1963. Ficus in the Pacific region. Pp. 233-45 in J.L. Gressitt (ed.), *Pacific Basin biogeography*. Tenth Pac. Sci. Congr., Bishop Museum Press, Hawaii.

_____ 1965. Checklist of Ficus in Asia and Australasia. *Gdns Bull., Singapore*, **21** : 1-186.

Couvée, L.M.J., Nugteren, D.H., and Luyken, R. 1962. The nutritional condition of the Kapaukus in the central highlands of Netherlands New Guinea. *Trop. Geogr. Med.*, **14** : 27-32.

Cranstone, B.A.L. 1961. *Melanesia: a short ethnography*. British Museum, London.

_____ 1972. Material culture. Pp. 715-40 in P. Ryan (ed.), *Encyclopaedia of Papua and New Guinea*, vol. II. Melbourne Univ. Press.

Darlington, P.J. 1957. *Zoogeography, the geographical distribution of animals*. Wiley, New York.

Daubenmire, R. 1972. Some ecologic consequences of converting forest to savanna in north-western Costa Rica. *Trop. Ecol.*, **13**(1): 31-51.

Dickie, J. and Malcolm, D.S. 1940. Note on a salt substitute used by one of the inland tribes of New Guinea. *J. Polynesian Soc.*, **49** : 144-7.

Diels, L. 1913. Die Annonaceen von Papuasien. *Bot. Jb.*, **49** : 113-67.

_____ 1930. Ein Beitrag zur Analyse der Hochgebirgsflora von Neu Guinea. *Bot. Jb.*, **63** : 324-9.

_____ 1934. Die Flora Australiens und Wegeners Verschiebungs Theorie. *Sber. preuss. Akad. Wiss.*, **33** : 533-45.

_____ 1936. The genetic phytogeography of the SW. Pacific area with particular reference to Australia. Pp. 189-94 in T.H. Goodspeed (ed.), *Essays in Geobotany*, University of California, Cambridge.

Eden, M.J. 1970. Savanna vegetation in the Northern Rupununi, Guyana. *J. trop. Geogr.*, **30** : 17-28.

Endacott, N.D. 1971. Implications of the wood chip industry as it affects the New Guinea scene. *Aust. Timb. J.*, **37**(4) : 54-61.

Engler, A. (revised by Melchior) 1954, 1964. *Syllabus der Pflanzenfamilien*, 12th ed., vols. I and II. Gebrüder Borntraeger, Berlin.

Fagan, R.H. and McAlpine, J.R. 1972. Population and land use of the Aitape-Ambunti area. *CSIRO Aust. Land Res. Ser.*, **30** : 126-32.

Feachem, R. 1973. The religious belief and ritual of the Raiapu Enga. *Oceania*, **43** : 259-85.

Flenley, J.R. 1967. The present and former vegetation of the Wabag region of New Guinea. Ph.D. thesis, Aust. Nat. Univ., Canberra.

_____ 1972. Evidence of Quaternary vegetational change in New Guinea. Pp. 99-108 in P. and M. Ashton (eds.), *The Quaternary era in Malesia*. Dept of Geog., Misc. Series 13, Univ. of Hull.

Floyd, A.G. 1954. Final report on the ethnobotanical expedition, West Nakanai, New Britian. Unpublished, Divn of Botany, Dept of Forests, Lae.

Fosberg, F.R. 1961. A classification of vegetation for general purposes. *Trop. Ecol.*, **2**(1-2) : 1-28.

Fox, J.E.D. 1970. Natural regeneration of the Kambui Hills forest in eastern Sierra Leone. Part I. Ecological status of the *Lophira/Heritiera* rain forest. *Trop. Ecol.*, **11**(2) : 169-85.

_____ and Tan Teong Hing 1971. Soils and forest on an ultrabasic hill north east of Ranau, Sabah. *J. trop. Geogr.*, **32** : 38-48.

Freund, A.P.H., Henty, E.E., and Lynch, M.A. 1965. Salt-making in inland New Guinea. *Trans. PNG scient. Soc.*, **6** : 16-19.

Galloway, R.W., Hope, G.S., Löffler, E., and Peterson, J.A. 1973. Late Quaternary glaciation and periglacial phenomena in Australia and New Guinea. Pp. 125-38 in E.M. van Zinderen Bakker (ed.), *Palaeoecology of Africa*, vol. VIII. Cape Town.

Gessner, F. 1967. Untersuchungen an der Mangrove in Ost-Venezuela. *Int. Revue ges. Hydrobiol. Hydrogr.*, **52**(5) : 769-81.

Gibbs, L.S. 1917. *A contribution to the phytogeography and flora of the Arfak mountains*. London.

Gillison, A.N. 1969. Plant succession in an irregularly fired grassland area — Doma Peaks region, Papua. *J. Ecol.*, **57** : 415-27.

_____ 1970. Structure and floristics of a montane grassland/forest transition, Doma Peaks region, Papua. *Blumea*, **18**(1) : 71-86.

_____ 1971. Dynamics of biotically induced grassland/forest transitions in Papua New Guinea. M.Sc. thesis, Aust. Nat. Univ., Canberra.

_____ 1972. The tractable grasslands of Papua New Guinea. Pp. 161-72 in *Change and development in rural Melanesia*. Fifth Waigani Seminar, Aust. Nat. Univ., Canberra, and Univ. of Papua New Guinea, Port Moresby.

Glick, L.B. 1967. Medicine as an ethnographic category: the Gimi of the New Guinea Highlands. *Ethnology*, **6** : 31-56.

_____ 1972. Medicine, indigenous. Pp. 756-7 in P. Ryan (ed.), *Encyclopaedia of Papua and New Guinea*. Melbourne Univ. Press.

Golson, J. 1974a. Archaeology and agricultural history in the New Guinea Highlands. In G. de G. Sieveking (ed.), *Essays in Economic Prehistory*. In press.

_____ 1974b. Recent discoveries in the New Guinea Highlands: simple tools and complex technology. Unpublished, Dept Prehistory, Aust. Nat. Univ., Canberra.

_____ Lampert, R.J., Wheeler, J.M., and Ambrose, W.R. 1967. A note on carbon dates for horticulture in the New Guinea Highlands. *J. Polynesian Soc.*, **76** : 369-71.

Good, R. 1960. On the geographical relationships of the angiosperm flora of New Guinea. *Bull. Br. Mus. nat. Hist. Botany*, **2** : 205-26.

_____ 1963. On the biological and physical relationships between New Guinea and Australia. Pp. 301-8 in J.L. Gressitt (ed.), *Pacific Basin biogeography*. Tenth Pac. Sci. Congr., Bishop Museum Press, Hawaii.

_____ 1964. *The geography of the flowering plants*, 3rd ed. Longmans Green, London.

Gorman, C.F. 1969. Hoabinhian: a pebble-tool complex with early plant associations in Southeast Asia. *Science*, **163** : 671-3.

_____ 1971. The Hoabinhian and after: subsistence patterns in SE Asia during the last Pleistocene and early Recent periods. *World Archaeology*, 2(3) : 300-20.

Gray, B. 1973. Distribution of *Araucaria* in Papua New Guinea. *Dept of For. Research Bull.*, **1**.

Green, P.S. 1969. Old World *Heliconia*. *Kew Bull.*, **23** : 471-8.

Gressitt, J.L. 1961. Problems in zoogeography of Pacific and Antarctic insects. *Pac. Ins. Monogr.*, **2** : 1-94.

Grubb, P.J., Lloyd, J.R., Pennington, T.D., and Whitmore, T.C. 1963. A comparison of montane and lowland rain forest in Ecuador. I. The forest structure, physiognomy, and floristics. *J. Ecol.*, **51** : 567-601.

_____ and Whitmore, T.C. 1966. A comparison of montane and lowland rain forest in Ecuador. II. The climate and its effects on the distribution and physiognomy of the forests. *J. Ecol.*, **54** : 303-33.

Haantjens, H.A., Mabbutt, J.A., and Pullen, R. 1965. Environmental influences in anthropogenic grasslands in the Sepik plains, New Guinea. *Pacif. Viewpoint*, 6(2) : 215-19.

Haddon, A.C. 1916. Kava-drinking in New Guinea. *Man*, **16** : 145-52.

_____ and Hornell, J. 1937. *Canoes of Oceania*, Vol. II : The Canoes of Melanesia, Queensland and New Guinea. Bishop Museum, Spec. Publ. 28, Honolulu.

Hamilton, L. 1955. Indigenous versus introduced vegetables in the village dietary. *P.N.G. Agric. J.*, **10** : 54-7.

Hartley, T.G. 1973. A survey of New Guinea plants for alkaloids. *Lloydia*, 36(3) : 217-319.

Hatanaka, S. and Bragge, L.W. 1973. Habitat isolation and subsistence economy in the central range of New Guinea. *Oceania,* **44**(1) : 38-57.

Havel, J.J. 1971. The araucaria forests of New Guinea and their regenerative capacity. *J. Ecol.*, **59**(1) : 203-14.

Heider, K.G. 1970. *The Dugum Dani: a Papuan culture in the highlands of West New Guinea*. Viking Fund Publ. in Anthrop. 49, Aldine, Chicago.

Held, G.J. 1957. *The Papuas of Waropen*. Martinus Nijhoff, The Hague.

Henry, T. and Muia, G. 1959. Special report on the diet of the Sepik River people. *Papua New Guin. agric. J.,* **12**(1) : 42-3.

Henty, E.E. 1960. Two drug plants in native culture. *Trans. PNG Scient. Soc.*, **1** : 19-20.

_____ 1969. *A manual of the grasses of New Guinea*. Botany Bull. 1, Dept of Forests, Papua New Guinea.

Heyligers, P.C. 1965. Vegetation and ecology of the Port Moresby-Kairuku area. *CSIRO Aust. Land Res. Ser.*, **14** : 146-73.

———— 1966. Observations on *Themeda australis-Eucalyptus* savannah in Papua. *Pacif. Sci.*, **20**(4) : 477-89.

———— 1967. Vegetation and ecology of Bougainville and Buka islands. *CSIRO Aust. Land Res. Ser.*, **20** : 121-45.

———— 1972a. Vegetation and ecology of the Aitape-Ambunti area. *CSIRO Aust. Land Res. Ser.*, **30** :73-99.

———— 1972b. Descriptions and ecology of the vegetation types of the Vanimo area. *CSIRO Aust. Land Res. Ser.*, **31** : 86-98.

———— 1972c. Analysis of the plant geography of the semideciduous scrub and forest and the eucalypt savannah near Port Moresby. *Pacif. Sci.*, **26** : 229-41.

Hide, R. 1974. Preliminary checklist of some plants in Nimai territory, Sinasina L.G.C. area, Chimbu District (1971-73). Unpublished.

Hipsley, E.H. and Clements, F.W. 1950. *Report of the N.G. Nutrition Survey Expedition 1947.* Govt Printer, Sydney.

———— and Kirk, N.E. 1965. *Studies of dietary intake and the expenditure of energy by New Guineans.* South Pac. Comm. Techn. Paper 147, Noumea.

Hodges, K., Fysh, C.F., and Rienits, K.G. 1950. New Guinea and Papuan food composition tables. Pp. 269-80 in E.H. Hipsley and F.W. Clements (eds.), *Report of the New Guinea Nutrition Survey Expedition 1947*, part 9. Govt Printer, Sydney.

Hogbin, H.I. 1938-9. Tillage and collection: a New Guinea economy. *Oceania*, **9** : 127-51, 286-325.

———— 1951. *Transformation scene: the changing culture of a New Guinea village.* Routledge and Kegan Paul, London.

Holdsworth, D.K. 1973. Some medicinal plants of the Marawaka Kukukuku people. *S.I.N.G.*, **1**(3-4) : 17-20.

———— 1974a. A phytochemical survey of medicinal plants in Papua New Guinea. Part I. *S.I.N.G.*, **2**(2) : 142-54.

———— 1974b. A phytochemical survey of medicinal plants of the d'Entrecasteaux Islands, Papua. *S.I.N.G.*, **2**(2) : 164-71.

———— and Farnworth, E.R. 1974. A phytochemical survey of medicinal and poisonous plants of the Central District of Papua. *S.I.N.G.* **2**(2) : 155-63.

———— and Heers, G. 1971. Some medicinal and poisonous plants from the Trobriand Islands, Milne Bay District. *Rec. Papua New Guinea Museum*, **1**(2) : 37-40.

———— and Longley, R.P. 1972. Some medicinal and poisonous plants from the Southern Highlands District of Papua. *Proc. Papua New Guinea scient. Soc.*, **24** : 21-4.

———— and N'Drawii, C.S. 1973. Medicinal and poisonous plants from Manus Island. *S.I.N.G.*, **1**(3-4) : 11-16.

———— and Tringen, S.B. 1973. Medicinal plants of the Sepik. *S.I.N.G.*, **1**(3-4) : 5-10.

Hoogland, R.D. 1951. Dilleniaceae. Pp. 141-74 in C.G.G.J. van Steenis (ed.), *Flora Malesiana,* vol. I(4). Jakarta.

———— 1958. The alpine flora of Mount Wilhelm (New Guinea). *Blumea.,* Suppl. 4(Dr H.J. Lam Jubilee Vol.): 220-38.

———— 1972. Plant distribution patterns across the Torres Strait. Pp. 131-52 in D. Walker (ed.), *Bridge and barrier : the natural and cultural history of Torres Strait.* Publ. BG/3, Aust. Nat. Univ., Canberra.

Hooker, H.D. 1860. Flora Tasmaniae I, Introductory essay: i-cxxviii.

Hope, G.S. 1973. The vegetation history of Mt. Wilhelm. Ph.D. thesis, Aust. Nat. Univ., Canberra.

_____ in press. The vegetation of the Jaya Mountains. Chapter 10 in G.S. Hope, J.A. Peterson and U. Radok (eds.), *Gunung Es — results of the Carstensz Glaciers Expeditions, Irian Jaya*.

Hope, J.H. and G.S. 1975. Palaeoenvironments for man in New Guinea. In R.L. Kirk and A.G. Thorne (eds.), *The biological origin of the Australians*. Inst. Aboriginal Studies, Canberra.

_____ and Peterson, J.A. 1975. Glaciation and vegetation in the high New Guinea mountains. Pp. 155-62 in R.P. Suggate and M.M. Creswell (eds.), *Quaternary studies*. R. Soc. N. Z. Bull. 13, Wellington.

Howlett, D.R. 1962. A decade of change in the Goroka Valley, New Guinea: land use and development in the 1950s. Ph.D. thesis, Aust. Nat. Univ., Canberra.

Isbell, R.F. 1969. The distribution of black spear grass (*Heteropogon contortus*) in tropical Queensland. *Tropical Grasslands*, 3(1) : 35-41.

Jack, W.H. 1961. The spatial distribution of tree stems in a tropical high forest. *Emp. For. Rev.*, **40**(3) : 234-41.

Jermy, A.C. 1965. *British Museum (Natural History) University of Newcastle-upon-Tyne Expedition to New Guinea*. London.

Johnson, L.A.S. and Briggs, B.G. 1963. Evolution in the Proteaceae. *Aust. J. Bot.*, **11** : 21-61.

Jones, E.W. 1955. Ecological studies on the rain forest of southern Nigeria. IV. The plateau forest of the Okumu Forest Reserve. *J. Ecol.*, **43** : 546-94.

Jones, W.T. 1971. The field identification and distribution of mangroves in eastern Australia. *Qd Nat.*, **20**(1-3) : 35-51.

Kaberry, P.M. 1941. The Abelam tribe, Sepik District, New Guinea: a preliminary report. *Oceania*, **11** : 233-58, 345-67.

Kalkman, C. 1955. A plant-geographical analysis of the Lesser Sunda Islands. *Acta bot. neerl.*, **4** : 200-25.

_____ and Vink, W. 1970. Botanical exploration in the Doma Peaks region, New Guinea. *Blumea*, **18**(1) : 87-135.

Keng, H. 1970. Size and affinities of the flora of the Malay Peninsula. *J. trop. Geogr.*, **41** : 43-56.

Kershaw, A.P. 1970. A pollen diagram from Lake Euramoo, north-east Queensland, Australia. *New Phytol.*, **69** : 785-805.

_____ 1971. A pollen diagram from Quincan Crater, north-east Queensland, Australia. *New Phytol.*, **70** : 669-81.

Koeppen, A. von and Cohen, W.E. 1955. Pulping studies of five species of a mangrove association. *Aust. J. appl. Sci.*, **6**(1) : 105-16.

Kooijman, S. 1962. Material aspects of the Star Mountain's culture. *Nova Guinea, Anthropology*, **10**(2) : 15-44.

Kruckeberg, A.R. 1969. Plant life on serpentinite and other ferromagnesian rocks in northwestern North America. *Syesis*, **2**(1-2) : 15-114.

Küchler, A.W. 1972. The mangrove in New Zealand. *N.Z. Geogr.*, **28**(2) : 113-29.

Kunkel, G. 1966. Über die Struktur und Sukzession der Mangrove Liberias und deren Randformationen. *Ber. schweiz. bot. Ges.*, **75** : 20-40.

Lam, H.J. 1934. Materials towards a study of the flora of the island of New Guinea. *Blumea*, **1** : 115-59.

_____ 1945. '*Fragmenta Papuana V*' Sargentia 5. Arnold Arboretum, Harvard University.

Landtman, G. 1933. *Ethnographical collection from the Kiwai district of British New Guinea, in the National Museum of Finland, Helsingfors (Helsinki); a descriptive survey of the material culture of the Kiwai people*. Commission of the Antell Collection, Helsinki.

Lane-Poole, C.E. 1925. *The forest resources of the Territories of Papua and New Guinea*. Govt Printer, Melbourne.

Lang, G.E., Knight, D.H., and Anderson, D.A. 1971. Sampling the density of tree species with quadrats in a species-rich tropical forest. *Forest Sci.*, **17**(3) : 395-400.

Latter, J.H. 1960. Aerial photographs as a guide to geological reconnaissance in the Solomon Islands. *Phot. Record*, **3** : 243-52.

Lawrence, P. 1972. Religion and Magic. Pp. 1001-12 in P. Ryan(ed.), *Encyclopaedia of Papua and New Guinea*, vol. II. Melbourne Univ. Press.

Lea, D.A.M. 1965. The Abelam: a study in local differentiation. *Pacif. Viewpoint*, **6**(2) : 191-214.

────── 1966. Yam growing in the Maprik area. *Papua New Guinea agric. J.*, **18** : 5-15.

────── 1970. Staple crops and main sources of food. Pp. 54-5 in R.G. Ward and D.A.M. Lea (eds.), *An atlas of Papua and New Guinea*. Collins-Longman, Glasgow.

────── 1972. Agriculture, indigenous. Pp. 10-18 in P. Ryan (ed.), *Encyclopaedia of Papua and New Guinea*, vol. I. Melbourne Univ. Press.

────── 1975. Human sustenance and the tropical forest. Paper presented to the MAB Conference, Port Moresby.

Lerche, G. and Steensberg, A. 1973. Observations on spade-cultivation in the New Guinea Highlands. *Tools and Tillage*, **2** : 27-104, 118.

Löffler, E. 1972. Pleistocene glaciation in Papua and New Guinea. *Z. Geomorph.*, Suppl. **13** : 32-58.

Luyken, R. and Luyken-Koning, F.W.M. 1955. Nutritional state of the Marind-Anim tribe in south New Guinea. *Documenta Med. geogr. trop.*, **7** : 315-39.

Luzbetak, L.J. 1958. Treatment of disease in the New Guinea Highlands. *Anthrop. Q.*, **31**(2) : 42-55.

Lyon, G.L., Peterson, P.J., Brooks, R.R., and Butler, G.W. 1971. Calcium, magnesium and trace elements in a New Zealand serpentine flora. *J. Ecol.*, **59**(2), 421-9.

Maahs, A.M. 1950. Salt-makers of the Wahgi. *Walkabout*, **16** : 15-18.

Macnae, W. 1966. Mangroves in eastern and southern Australia. *Aust. J. Bot.*, **14**(1) : 67-104.

────── and Kalk, M. 1962. The ecology of the mangrove swamps at Inhaca Island, Moçambique. *J. Ecol.*, **50** : 19-34.

Malinowski, B. 1918. Fishing in the Trobriand Islands. *Man*, **18** : 87-92.

────── 1935. *Coral gardens and their magic*, 2 vols. Allen and Unwin, London.

Mann, G. 1968. Die Ökosysteme Südamerikas. Pp. 171-229 in E.J. Fittkau, J. Illies, H. Klinge, G.H. Schwabe, H. Sioli(eds.), *Biogeography and ecology in South America*, Vol. I. W. Junk, The Hague.

Massal, E. and Barrau, J. 1956. *Food plants of the South Sea Islands*. South Pac. Comm. Techn. Paper 94, Noumea.

Mattfeld, J. 1929. Die Compositien von Papuasien. *Bot. Jb.*, **62** : 386-451, 494-501.

McAlpine, J.R. 1967. Population and land use of Bougainville and Buka Islands. *CSIRO Aust. Land Res. Ser.*, **20** : 157-67.

────── 1968. Population, land use, and transport in the Wewak-Lower Sepik area. *CSIRO Aust. Land Res. Ser.*, **22** : 133-40.

Meggitt, M.J. 1957. House building among the Mae Enga, Western Highlands, Territory of New Guinea. *Oceania*, **27** : 161-76.

────── 1958a. The Enga of the New Guinea Highlands: some preliminary observations. *Oceania*, **28** : 253-330.

———— 1958b. Salt manufacture and trading in the Western Highlands of New Guinea. *The Australian Museum Magazine*, 12 : 309-13.

Merrill, E.D. 1926. *An enumeration of Philippine flowering plants*, vol. IV: 77-104. Bureau of Science, Manila.

Mervart, J. 1972. Frequency curves of the growing stock in the Nigerian high forests. *Nigerian J. For.*, **2**(1) : 7-15.

Miklouho-Maclay, N. de 1886. List of plants in use by the natives of the Maclay coast, New Guinea. *Proc. Linn. Soc. N.S.W.*, **10** : 346-58.

Montgomery, D.E. 1960. Patrol of the Upper Chimbu Census Division, Eastern Highlands. *Papua New Guin. agric. J.*, **13** : 1-9.

Muller, J. 1972. Palynological evidence for change in geomorphology, climate and vegetation in the Mio-Pliocene of Malesia. Pp. 6-16 in P. and M. Ashton (eds.), *The Quaternary era in Malesia*. Dept of Geog., Misc. Series 13, Univ. of Hull.

Murdock, G.P. 1967. *Ethnographic atlas*. Univ. of Pittsburgh Press.

Newton, D. 1972. Art. Pp. 29-50 in P. Ryan (ed.), *Encyclopaedia of Papua and New Guinea*, vol. I. Melbourne Univ. Press.

Nilles, J. 1942-5. Digging sticks, spades, hoes, axes and adzes of the Kuman people in the Bismarck Mountains of East-Central New Guinea. *Anthropos*, **37-40** : 205-12.

Nix, H.A. and Kalma, J.D. 1972. Climate as a dominant control in the biogeography of northern Australia and New Guinea. Pp. 61-92 in D. Walker (ed.), *Bridge and barrier: the natural and cultural history of Torres Strait*. Pub. BG/3, Aust. Nat. Univ., Canberra.

Oomen, H.A.P.C. 1971. Ecology of human nutrition in New Guinea: evaluation of subsistence patterns. *Ecol. Food and Nutrit.*, **1** : 1-16.

———— and Corden, M.W. 1970. *Metabolic studies in New Guineans: nitrogen metabolism in sweet potato eaters*. South Pac. Comm. Techn. Paper 163, Noumea.

———— and Malcolm, S.H. 1958. *Nutrition and the Papuan child*. South Pac. Comm. Tech. Paper 118, Noumea.

Oosterwal, G. 1961. *People of the Tor*. Van Gorcum, Assen.

Paijmans, K. 1966. Typing of tropical vegetation by aerial photographs and field sampling in northern Papua. *Photogrammetria*, **21** : 1-25.

———— 1967. Vegetation of the Safia-Pongani area. *CSIRO Aust. Land Res. Ser.*, **17** : 142-67.

———— 1969. Vegetation and ecology of the Kerema-Vailala area. *CSIRO Aust. Land Res. Ser.*, **23** : 95-116.

———— 1970. Land evaluation by air photo interpretation and field sampling in Australian New Guinea. *Photogrammetria*, **26**(2/3) : 77-100.

———— 1971. Vegetation, forest resources, and ecology of the Morehead-Kiunga area. *CSIRO Aust. Land Res. Ser.*, **29** : 88-113.

———— 1973a. Plant succession on Pago and Witori volcanoes, New Britain. *Pacif. Sci.*, **27**(3) : 260-8.

———— 1973b. Vegetation of Eastern Papua. *CSIRO Aust. Land Res. Ser.*, **32** : 89-125.

———— 1975. Explanatory notes to the vegetation map of Papua New Guinea. *CSIRO Aust. Land Res. Ser.*, **35**.

———— and Löffler, E. 1972. High-altitude forests and grasslands of Mt. Albert Edward, New Guinea. *J. trop. Geogr.*, **34** : 58-64.

Panoff, F. 1970a. Maenge remedies and conception of disease. *Ethnology*, **9**(1) : 68-84.

———— 1970b. A feminine costume in New Britain. *J. Polynesian Soc.*, **79** : 99-106.

———— 1972. Maenge gardens: a study of Maenge relationship to domesticates. Ph.D. thesis, Aust. Nat. Univ., Canberra.

Papua New Guinea Department of Forests 1973. *New horizons — forestry in Papua New Guinea.* Jacaranda, Brisbane.

Pedley, L. and Isbell, R.F. 1971. Plant communities of Cape York Peninsula. *Proc. R. Soc. Qd*, **82**(5) : 51-74.

Peters, F.E. 1958. *The chemical composition of South Pacific foods.* South Pac. Comm. Techn. Paper 115, Noumea.

Peterson, J.A. and Hope, G.S. 1972. Lower limit and maximum age for the last major advance of the Carstensz Glaciers, West Irian. *Nature Lond.*, **240**(5375) : 36-7.

Phillips, F.H. and Watson, A.J. 1959. *Pulping studies on New Guinea woods, I.* CSIRO Aust. Div. For. Prod. Techn. Paper 8.

Poore, M.E.D. 1964. Integration in the plant community. *J. Ecol.*, Suppl. **52** : 213-26.

―――― 1968. Studies in Malaysian rain forest. I. The forest on Triassic sediments in Jengka Forest Reserve. *J. Ecol.*, **56**(1) : 143-96.

Pospisil, L. 1963. *Kapauku Papuan economy.* Publ. 67, Dept of Anthrop., Yale Univ., New Haven.

Powell, J.M. 1970a. The impact of man on the vegetation of the Mt Hagen region, New Guinea. Ph.D. thesis, Aust. Nat. Univ., Canberra.

―――― 1970b. The history of agriculture in the New Guinea Highlands. *Search*, **1** : 199-200.

―――― 1973. Agricultural practices of the Huli of Papua. Unpublished.

―――― 1974a. A note on wooden gardening implements of the Mt Hagen region, New Guinea. *Records Papua New Guinea Museum*, **4** : 21-8.

―――― 1974b. Traditional legumes of the New Guinea Highlands. *S.I.N.G.*, **2**(1) : 48-62.

―――― , Kulunga, A., Moge, R., Pono, C., Zimike, F., and Golson, J. 1975. *Agricultural traditions of the Mt Hagen area.* Geog. Dept Occas. Paper 12, Univ. PNG, Port Moresby.

Proctor, J. 1971a. The plant ecology of serpentine. II. Plant response to serpentine soils. *J. Ecol.*, **59**(2) : 397-410.

―――― 1971b. The plant ecology of serpentine. III. The influence of a high magnesium/calcium ratio and high nickel and chromium levels in some British and Swedish serpentine soils. *J. Ecol.*, **59**(3) : 827-42.

―――― and Woodell, S.R.J. 1971. The plant ecology of serpentine. I. Serpentine vegetation of England and Scotland. *J. Ecol.*, **59**(2) : 375-95.

Purseglove, J.W. 1968. *Tropical Crops: Dicotyledons*, 2 vols. Longmans Green, London.

―――― 1972. *Tropical Crops: Monocotyledons*, 2 vols. Longmans Green, London.

Rand, A.L. and Brass, L.J. 1940. Results of the Archbold Expeditions. No. 29 Summary of the 1936-1937 New Guinea Expedition. *Bull. Am. Mus. nat. Hist.*, **77** : 341-80.

Rappaport, R.A. 1967. *Pigs for the ancestors : ritual in the ecology of a New Guinea people.* Yale Univ. Press, New Haven.

Raven, P.H. and Axelrod, D.I. 1972. Plate tectonics and Australasian paleobiogeography. *Science N.Y.*, **176** : 1379-86.

Reiner, E.J. and Robbins, R.G. 1964. The middle Sepik plains, New Guinea. A physiographic study. *Geogrl Rev.*, **54**(1) : 20-44.

Reynders, J.J. 1962. Shifting cultivation in the Star Mountains area. *Nova Guinea*, Anthropology, **10**(3) : 45-73.

Richards, P.W. 1963. Ecological notes on West African vegetation. II. Lowland forest of the Southern Bakundu Forest Reserve. *J. Ecol.*, **51** : 123-49.

―――― 1964. *The tropical rain forest*, 3rd ed. Cambridge Univ. Press.

Richardson, S.D. 1970. Mission to West Irian. *Commonw. For. Rev.*, **49**(2) : 138-49.

Riesenfeld, A. 1946. Rattan cuirasses and gourd penis-cases in New Guinea. *Man*, **46** : 31-6.
_____ 1951. Tobacco in New Guinea and other areas of Melanesia. *Jl R. anthrop. Inst. Gt. Brit. and Ireland*, **81** : 69-102.
Riley, E.B. 1923. Sago-making on the Fly River. *Man*, **23** : 145-6.
Robbins, R.G. 1959. Vegetation of the lower Ramu-Atitau area. *CSIRO Aust. Land Res. Div. Report* **59/1** : 99-118.
_____ 1962. The podocarp-broadleaf forests of New Zealand. *Trans. R. Soc. N.Z.*, *Botany*, **1**(5) : 33-75.
_____ 1963. The anthropogenic grasslands of Papua and New Guinea. Pp. 313-99 in *Proc. Symp. on 'The impact of man on humid tropics vegetation'*, *Goroka 1960*. UNESCO, Djakarta.
_____ 1968. Vegetation of the Wewak-Lower Sepik area. *CSIRO Aust. Land Res. Ser.*, **22** : 109-24.
_____ 1969. A prerequisite to understanding tropical rainforest. *Malay. Forester*, **32**(4) : 361-3.
_____ 1970. Vegetation of the Goroka-Mount Hagen area. *CSIRO Aust. Land Res. Ser.*, **27** : 104-18.
_____ and Pullen, R. 1965. Vegetation of the Wabag-Tari area. *CSIRO Aust. Land Res. Ser.*, **15** : 100-15.
Rollet, B. 1968. Étude quantitative de profils structuraux de forêts denses Vénézuéliennes. Comparaison avec d'autres profils de forêts denses tropicales de plaine. *Adansonia*, Ser. 2, **8**(4) : 523-49.
Royen, P. van 1959. Compilation of keys to the families and genera of Angiosperms and Gymnosperms in New Guinea. Unpublished.
_____ 1963a. Sertulum Papuanum 7. Notes on the vegetation of South New Guinea. *Nova Guinea Bot.*, **13** : 195-241.
_____ 1963b. Sertulum Papuanum 9. Blanche Bay's vulcanoseres. *Trans. Papua New Guinea Sci. Soc.*, **4** : 3-9.
Ruinard, J. 1967. Notes on sweet potato research in West New Guinea (West Irian). Pp. 88-111 in E.A. Tai *et al.* (eds.), *Proceedings International Symposium on Tropical Root Crops*, vol. I(3). Univ. of West Indies Press, Trinidad.
Salisbury, R.F. 1961. *From stone to steel*. Melbourne Univ. Press.
Saunders, J.C. 1957. Vegetation of the Gogol-Upper Ramu area. *CSIRO Aust. Land Res. Div. Report* **57/2** : 57-66.
Schiefenhoevel, W. 1971. Aspects of the medical system of the Kaluli and Waragu Language-Group, Southern Highlands District. *Mankind*, **8** : 141-5.
Schindler, A.J. 1952. Land use by natives of Aiyura village, Central Highlands, New Guinea. *South Pacific*, **6**(2) : 302-7.
Schodde, R. 1972. A review of the family Pittosporaceae in Papuasia. *Aust. J. Bot.*, Suppl. Ser., **3** : 3-60.
_____ and Calaby, J.H. 1972. The biogeography of the Australo-Papuan bird and mammal faunas in relation to Torres Strait. Pp. 257-300 in D. Walker (ed.), *Bridge and barrier : the natural and cultural history of Torres Strait*. Publ. BG/3, Aust. Nat. Univ., Canberra.
Schulz, J.P. 1960. *Ecological studies on rain forest in northern Suriname*. Amsterdam.
Serpenti, L.M. 1965. *Cultivators in the swamps*: social structure and horticulture in a New Guinea society. Assen.
Simmonds, N.W. 1959. *Bananas*. Longmans, London.
Sleumer, H.O. 1966. Ericaceae. Pp. 469-914 in C.G.G.J. van Steenis (ed.), *Flora Malesiana*, vol. I(6). Noordhoff, Groningen.
Smith, A.C. 1943. Taxonomic notes on the Old World species of Winteraceae. *J. Arnold Arbor.*, **24** : 119-64.

―――― 1945. Geographical distribution of the Winteraceae. *J. Arnold Arbor.*, **26** : 48-59.

Smith, A.C. 1970. *The Pacific as a key to flowering plant history.* H.L. Lyon Arboretum Lecture No. 1.

―――― Briden, J.C., and Drewry, G.E. 1973. Phanerozoic world maps. Pp. 1-39 in N.F. Hughes (ed.), *Organisms and continents through time.* Spec. Pap. Palaeont. 12, Palaeontological Assoc., London.

Sorenson, E.R. and Gajdusek, D.C. 1969. Nutrition in the Kuru Region I. Gardening, food handling and diet of the Fore People. *Acta trop.*, **26** : 281-330.

Stanhope, J.M. 1968. Competing systems of medicine among the Rao-Breri, Lower Ramu River, New Guinea. *Oceania*, **39** : 137-45.

Steenis, C.G.G.J. van 1934-6. On the origin of the Malaysian mountain flora. *Bull. Jard. bot. Buitenz. III*, **13** : 135-262, 289-417, **14** : 56-72.

―――― 1950. The delimitation of Malaysia and its main plant geographical divisions. Pp. lxx-lxxv in C.G.G.J. van Steenis (ed.), *Flora Malesiana*, vol. I(1). Noordhoff, Jakarta.

―――― 1952. Rheophytes. *Proc. R. Soc. Qd*, **62**(6) : 61-8.

―――― 1953. Papuan Nothofagus. *J. Arnold Arbor.*, **34** : 301-74.

―――― 1954. Vegetatie en flora. Pp. 218-75 in W.C. Klein (ed.), *Nieuw Guinea. De ontwikkeling op economisch, sociaal en cultureel gebied*, vol. II, *Nederlands en Australisch Nieuw Guinea*. Staatsdrukkerij-en uitgeverijbedrijf, The Hague.

―――― 1962a. The landbridge theory in Botany. *Blumea*, **11** : 235-542.

―――― 1962b. The mountain flora of the Malaysian tropics. *Endeavour*, **21**(83/84) : 183-93.

―――― 1965. Concise plant-geography of Java. Pp. 1-72 in C.A. Backer and R.C. Bakhuizen van den Brink (eds.), *Flora of Java*. Noordhoff, Groningen.

―――― 1968. Frost in the tropics. Pp. 154-67 in *Proc. Symp. Recent Advances in Tropical Ecology*, Part I. Faridabad, India.

―――― 1971. Nothofagus, key genus of plant geography, in time and space, living and fossil, ecology and phylogeny. *Blumea*, **19**(1) : 65-98.

―――― and Balgooy, M.M.J. van 1966. Pacific plant areas 2. *Blumea* Suppl. **5**.

Steenis-Kruseman, M.J. van 1950. Malaysian plant collectors and collections. Pp. 1-639 in C.G.G.J. van Steenis (ed.), *Flora Malesiana*, vol. I. Noordhoff, Jakarta.

―――― 1958. Malaysian plant collectors and collections. Pp. ccli-cccxxxvi in C.G.G.J. van Steenis (ed.), *Flora Malesiana*, vol. V. Noordhoff, Jakarta.

Stopp, K. 1963. Medicinal plants of the Mt Hagen people (Mbowamb) in New Guinea. *Econ. Bot.* **17**(1) : 16-22.

Story, R. 1970. Vegetation of the Mitchell-Normanby area. *CSIRO Aust. Land Res. Ser.*, **26** : 75-88.

Straatmans, W. 1967. Ethnobotany of New Guinea in its ecological perspective. *J. Agric. trop. Bot. appl.*, **14** : 1-20.

―――― 1971. An ethno-botanical checklist of New Guinea. Unpubl. mimeo. Aust. Nat. Univ., Canberra.

Strathern, A. and Strathern, A.M. 1971. *Self-decoration in Mount Hagen*. Duckworth, London.

Strathern, A.M. 1969. Why is the *Pueraria* a sweet potato? *Ethnology*, **8** : 189-98.

Taylor, B.W. 1957 Plant succession on recent volcanoes in Papua. *J. Ecol.*, **45** : 233-43.

―――― 1959. The classification of lowland swamp communities in northeastern Papua. *Ecology*, **40**(4) : 703-11.

——— 1963. An outline of the vegetation of Nicaragua. *J. Ecol.*, **51**(1) : 27-54.

——— 1964a. Vegetation of the Buna-Kokoda area. *CSIRO Aust. Land Res. Ser.*, **10** : 89-98.

——— 1964b. Vegetation of the Wanigela-Cape Vogel area. *CSIRO Aust. Land Res. Ser.*, **12** : 69-83.

Thorne, R.F. 1963. Biotic distribution patterns in the tropical Pacific. Pp. 311-54 in J.L. Gressitt (ed.), *Pacific basin biogeography.* Tenth Pac. Sci. Congr., Bishop Museum Press, Hawaii.

——— 1968. Synopsis of a putatively phylogenetic classification of the flowering plants. *Aliso*, **6** : 57-66.

Todd, J.A. 1934. Report on research work in south-west New Britian, Territory of New Guinea. *Oceania*, **5** : 193-213.

Tothill, J.C. 1969. Soil temperatures and seed burial in relation to the performance of *Heteropogon contortus* and *Themeda australis* in burnt native woodland pastures in eastern Queensland. *Aust. J. Bot.*, **17** : 269-75.

Townsend, P.K. 1971. New Guinea sago gatherers — a study of demography in relation to subsistence. *Ecol. Food and nutrition*, **1** : 19-24.

Treide, B. 1967. *Wildpflanzen in der Ernahrung der Grundbevölkerung Melanesiens.* Akademie-Verlag, Berlin.

Troll, C. 1957. Tropical mountain vegetation. Pp. 37-46 in *Proc. 9th Pacific Sci. Congress*, vol. xx. Science Soc. of Thailand, Bangkok.

——— 1959. Die tropischen Gebirge — Ihre dreidimensionale klimatische und pflanzengeographische Zonierung. *Bonn. geogr. Abh.*, **25** : 1-93.

Vega, L.C. 1968. La estructura y composición de los bosques húmedos tropicales del Carare, Colombia. *Turrialba*, **18**(4) : 416-36.

Venkatachalam, P.S. 1962. *A study of the diet, nutrition and health of the people of the Chimbu area.* Dept. Public Health, TPNG, Monography 4.

Waddell, E.W. 1972. *The mound builders: agricultural practices, environment, and society in the Central Highlands of New Guinea.* Univ. Washington Press, Seattle.

Wade, L.K. and McVean, D.N. 1969. *Mt Wilhelm Studies I: The alpine and subalpine vegetation.* Publ. BG/1, Aust. Nat. Univ., Canberra.

Walker, D. 1966. Vegetation of the Lake Ipea region, New Guinea Highlands. I. Forest, grassland and 'garden'. *J. Ecol.*, **54** : 503-33.

——— 1968. A reconnaissance of the non-arboreal vegetation of the Pindaunde catchment, Mount Wilhelm, New Guinea. *J. Ecol.*, **56** : 445-66.

——— 1970. The changing vegetation of the montane tropics. *Search*, **1** : 217-21.

——— 1972. Bridge and barrier. Pp. 399-405 in D. Walker (ed.), *Bridge and barrier: the natural and cultural history of Torres Strait.* Publ. BG/3, Aust. Nat. Univ., Canberra.

——— 1973. Highlands vegetation. *Aust. nat. Hist.*, **17**(12) : 410-14, 419.

Warburg, O. 1891. Beitrage zur Kenntnis der papuanischen Flora. *Bot. Jb.*, **13** : 230-455.

Wardle, P. 1971. An explanation for alpine timberline. *N.Z. Jl Bot.*, **9**(3) : 371-402.

Watson, J.B. 1964. A previously unreported root crop from the New Guinea Highlands. *Ethnology*, **3** : 1-5.

——— 1967. Horticultural traditions in the eastern New Guinea Highlands. *Oceania*, **38** : 81-98.

——— 1968. Pueraria: names and traditions of a lesser crop of the Central Highlands, New Guinea. *Ethnology*, **7** : 268-79.

Watson, J.G. 1928. *Mangrove forests of the Malay Peninsula.* Malay. Forest Rec. 6.

Webb, L.J. 1955. A preliminary phytochemical survey of Papua-New Guinea. *Pacif. Science*, **9** : 430-41.

─────── 1959. Some new records of medicinal plants used by the Aborigines of Tropical Queensland and New Guinea. *Proc. R. Soc. Qd*, **71** : 103-10.

─────── and Tracey, J.G. 1972. An ecological comparison of vegetation communities on each side of Torres Strait. Pp. 109-29 in D. Walker (ed.), *Bridge and barrier: the natural and cultural history of Torres Strait*. Publ. BG/3, Aust. Nat. Univ., Canberra.

Wedgwood, C.H. 1934. Report on research in Manam Island, Mandated Territory of New Guinea. *Oceania*, **4** : 373-403.

Whitaker, T.W. 1971. Endemism and pre-Columbian migration of the bottle gourd, *Lagenaria siceraria* (Mol.) Standl. Pp. 320-7 in C.L. Riley *et al.* (eds.), *Man across the sea*. Univ. Texas Press, Austin.

White, J.P. 1971. New Guinea: the first phase in Oceanic settlement. Pp. 45-52 in R.C. Green and M. Kelly (eds.), *Studies in Oceanic culture history 2*. Pac. Anthrop. Records 12, Bishop Museum Press, Honolulu.

─────── 1972. *Ol Tumbuna: archaeological excavations in the Eastern Central Highlands, Papua New Guinea*. Terra Australis 2, Dept Prehistory, Aust. Nat. Univ., Canberra.

─────── Crook, K.A.W. and Ruxton, B.P. 1970. Kosipe: a late Pleistocene site in the Papuan highlands. *Proc. prehist. Soc.*, **36** : 152-70.

White, K.J. 1971. The lowland rain forest in Papua New Guinea. Paper presented at Pacific Sc. Assoc. Pre Congress Conf. Bogor.

Whiting, J.W.M. and Reed, S.W. 1938. Kwoma culture: report on field work in the mandated Territory of New Guinea. *Oceania*, **9** : 170-216.

Whitmore, T.C. 1966a. *Guide to the forests of the British Solomon Islands*. Oxford Univ. Press, London.

─────── 1966b. The social status of *Agathis* in a rain forest in Melanesia. *J. Ecol.*, **54** : 285-301.

─────── 1969. Geography of the flowering plants (of the Solomons). *Phil. Trans. R. Soc.*, Ser. B, **255** : 549-66.

─────── 1973. Plate tectonics and some aspects of Pacific plant geography. *New Phytol.*, **72** : 1185-90.

─────── 1974. *Change with time and the role of cyclones in tropical rain forest on Kolombangara, Solomon Islands*. Comm. For. Inst. Paper 46, Univ. of Oxford.

Williams, F.E. 1930. *Orokaiva society*. Oxford Univ. Press, London.

─────── 1936. *Papuans of the Trans-Fly*. Clarendon Press, Oxford.

─────── 1940a. Natives of Lake Kutubu, Papua. *Oceania*, **11**(2) : 121-57.

─────── 1940b. *Drama of Orokolo*. Clarendon Press, Oxford.

Williams, P.W., McDougall, I., and Powell, J.M. 1972. Aspects of the Quaternary geology of the Tari-Koroba area, Papua. *J. geol. Soc. Aust.*, **18** : 333-47.

Williamson, R.W. 1912. *The Mafulu*. Macmillan, London.

Willis, J.C. (revised by H.K. Airy Shaw) 1973. *A dictionary of the flowering plants and ferns* 8th ed. Cambridge Univ. Press.

Wit, H.C.D. de 1949. Short history of the phytography of Malaysian vascular plants. Pp. lxx-clxi in C.G.G.J. van Steenis (ed.), *Flora Malesiana*, vol. IV. Noordhoff, Jakarta.

Womersley, J.S. 1958. The *Araucaria* forests of New Guinea. A unique vegetation type in Malaysia. Pp. 252-7 in *Proc. Symp. on Humid Trop. Veg.* Tjiawi (Indonesia).

─────── 1972. Crop Plants. Pp. 222-32 in P. Ryan (ed.), *Encyclopaedia of Papua and New Guinea*, vol. I Melbourne Univ. Press.

Womersley, J.S. 1973. Records of medicinal plants — Divn. Botany, Dept. of Forests, Lae. Paper presented at SPC Medicinal Plants Conference, Tahiti.

Wong, Y.K. and Whitmore, T.C. 1970. On the influence of soil properties on species distribution in a Malayan lowland dipterocarp rain forest. *Malay. Forester*, **33** : 42-54.

Wyatt-Smith, J. 1964. A preliminary vegetation map of Malaya with descriptions of the vegetation types. *J. trop. Geogr.,* **18** : 200-13.

Yen, D.E. 1963. Sweep potato variation and its relation to human migration in the Pacific. Pp. 93-117 in J. Barrau (ed.), *Plants and the migration of Pacific peoples*. Bishop Museum Press, Honolulu.

_____ 1968. Natural and human selection in the Pacific sweet potato. Pp. 387-412 in E.T. Drake (ed.), *Evolution and environment.* Yale Univ. Press, New Haven.

_____ 1971a. Construction of the hypothesis for distribution of the sweet potato. Pp. 328-42 in C.L. Riley *et al.* (eds.), *Man across the sea.* Univ. Texas Press, Austin.

_____ 1971b. The development of agriculture in Oceania, in R.C. Green and M. Kelly (eds.), *Studies in Oceanic culture history 2.* Pac. Anthrop. Rec. 12, Biship Museum Press, Honolulu.

_____ 1972. Ethnobotany. Pp. 380-4 in P. Ryan (ed.), *Encyclopaedia of Papua and New Guinea*, vol. I. Melbourne Univ. Press.

_____ 1973. The origins of Oceanic agriculture. *Arch. Phys. Anthrop. in Oceania*, **8** : 68-85.

_____ 1974. *The sweet potato and Oceania: an essay in ethnobotany.* B.P. Bishop Museum Bulletin 236, Honolulu.

_____ and Wheeler, J.M. 1968. Introduction of taro into the Pacific: the indications of the chromosome numbers. *Ethnology*, **7**(3) : 259-67.

Index of Botanical Names

Abroma angusta, 108, 156, 165, 169
Abrotanella, 19
Abrus, 151
Acacia, 30, 35, 42, 48, 52, 53·
   leptocarpa, 53, 58
Acaena, 21
Acalypha, 108, 133, 148, 151, 152, 156, 160, 162
   insulana, 108, 133, 134, 148, 151, 152, 162
   wilkesiana, 146
Acanthaceae, 6, 131, 174
Acanthus ilicifolius, 29, 34
Aceraceae, 6
Acianthus, 21
Acorus calamus, 147, 148, 150
Acronychia, 152, 154
Acrostichum aureum, 29, 32, 33, 34, 35, 36
Acsmithia, 20
Adenanthera pavonina, 66, 144, 173
Adenia, 165
Adenostemma hirsutum, 148
Aegialites, 7
Aegialitidaceae, 7
Aegiceras, 11
   corniculatum, 34
Aeschynanthus, 20, 165
Agapetes, 20
Agathis, 104
   macrophylla, 70
Agave, 169
Ageratum, 96, 137, 180
   conyzoides, 139, 140, 144
Aglaia, 108, 114
   sapindina, 156, 162
Aglaonema, 148
Agropyron, 134
Agrostis, 21
   reinwardtii, 91, 94, 101
Aizoaceae, *see* Tetragoniaceae
Akaniaceae, 6
Albizia, 84, 108, 114, 148, 150, 151, 152
   falcataria, 61, 83, 144, 155, 158
   procera, 56, 74, 76, 78
Alectryon, 152
Aleurites, 151, 174
   moluccana, 108, 132, 151
Allium, 108, 116, 117, 133
Allophylus cobbe, 29, 156
Alloteropsis semialata, 57
Alocasia hollrungii, 148, 167

   macrorrhiza, 108, 166, 167
Alphitonia, 79, 139
   incana, 140, 141, 148, 162, 172
   moluccana, 140, 162
Alpinia, 108, 133, 136, 137, 139, 140, 141, 148, 162, 164, 166, 167, 173
Alseuosmiaceae, 7, 8
Alsophila, 116, 117
   glauca, see Cyathea tenggerensis
Alstonia, 143, 158, 162
   brassii, 140
   scholaris, 44, 50, 51, 140, 143, 144, 145, 146, 147
   spatulata, 44
   spectabilis, 143, 146
Althoffia, 59, 61
   pleiostigma, 158, 169
Altingiaceae, 6
Amaracarpus, 21, 99
   brassii, 148
Amaranthus, 114, 116, 117, 119, 124, 129, 130, 168, 175, 180
   hybridus s.l., 108, 124
   tricolor, 108, 124
   viridis, 108, 124
Amborellaceae, 7
Amomum, 108, 133
   polycarpum, 108, 132, 148, 150
Amorphophallus campanulatus, 144
Ananas comosus, 115, 116, 117
Ancardiaceae, 174; *see also* Blepharocaryaceae, Pistaciaceae
Anaphalis, 19, 92
Anarthriaceae, 6
Ancistrocladaceae, 6
Angiopteris, 148
   evecta, 139
Anisophylleaceae, 6
Anisoptera polyandra, 52, 64, 66, 69, 84
Annesijoa, 11, 14
Annona, 124
   muricata, 108, 116, 117
Annonaceae, 2, 18, 86
Anthocephalus chinensis, 51
Anthoxanthum, 21
   angustum, 91, 94, 100
Antidesma, 160, 162
   ghaesembilla, 56, 78
Antoniaceae, 6
Aphanamixis, 162
Apiaceae, 19
Apluda mutica, 55, 57, 75

Apocynaceae, 143
Aquifoliaceae, *see* Sphenostemonaceae
Araceae, 107, 174
Aralia, 104
Araliaceae, 19, 86
Araucaria, 4, 64, 65, 70, 74, 81, 84, 89, 91, 104, 151
   cunninghamii, 70, 91, 152, 155, 157
   hunsteinii, 63, 68, 69, 70, 104, 148
Archboldiodendron, 22
Archidendron, 11, 135
Archontophoenix, 116, 117, 125, 135, 160, 162, 166
Ardisia, 148, 152, 162
Areca, 81, 151, 152, 155, 181
   catechu, 135, 141, 144, 146, 148, 152, 158, 159, 166, 173, 174, 181
Arenga, 108, 114, 147, 181
   pinnata, 181
   saccharifera, 135, 158, 159
Aristolochia, 142, 144, 146, 147
Arrhenechthites, 19, 108, 148
Arthraxon ciliaris, 91, 96, 162, 164
Arthropteris obliterata, 173
Artocarpus, 61, 108, 114, 131, 137, 158
   altilis, 59, 81, 108, 116, 117, 123, 132, 140, 144, 148, 150, 151, 156, 168, 169, 175
Arundinella furva, 91, 94
   setosa, 75
Ascarina, 11, 13, 20
   philippinensis, 108, 162, 172
Asplenium, 108, 125, 133
   affine, 108
   nidus, 133
Astelia, 11, 20, 102
   papuana, 101, 105, 147
Asteraceae, 2, 19, 174
Astilbe, 21, 108, 148
Astronia, 86, 87, 108, 133, 148, 162, 167, 172
Atherospermataceae, 7
Athyrium, 125, 137
   esculentum, 108, 116, 117
Atylosia, 76
Austrobaileyaceae, 6
Averrhoa carambola, 143
Averrhoaceae, 7
Avicennia, 32, 34, 35
   marina, 28, 32, 34
Azolla imbricata, 37, 38

201

Balanopaceae, 6
Balanophora, 11, 139
Balsaminaceae, 7
Bambusa, 108, 116, 117, 132, 148, 152, 155, 158, 162, 164, 173, 175, 181
  forbesii, 160, 162
Banksia dentata, 53, 54, 58, 78
Barclayaceae, 7
Barringtonia, 44, 108, 122, 124, 139, 144, 146, 157, 177
  asiatica, 29, 166, 167
  edulis, 124
  racemosa, 160, 162
  tetraptera, 41, 42, 48
Bassia, see *Madhuca*
Batidaceae, 7
Batis, 4, 9, 10
  argillicola, 4, 34
Baueraceae, 6
Bedfordia, see *Senecio*
Begonia, 65, 108, 133, 140
  angustae, 139
Begoniaceae, 7
Beilschmiedia, 108, 133, 148
Berberidaceae, 6
Betulaceae, 6
Bidens, 108, 133, 137, 138, 142, 151
  pilosa, 137
Bikkia, 11, 13
Bischofia javanica, 44, 51, 61
Bixa orellana, 143, 151
Blechnum, 108, 133
Blepharocarya, 18
Blepharocaryaceae, 6
Blumea, 137, 141, 144
  riparia, 108, 148, 165
Boehmeria, 140, 144, 168, 169
Boerlagiodendron, 133, 162
Bombax ceiba, 52, 66, 81
Bonettiaceae, 7
Boraginaceae, 20
Boronieae, 18
Borreria laevis, 141, 148
Brachychiton carruthersii, 66
Brachycome, 19
Brachyonostylum, 19
Brachypodium, 21
Brassica, 116, 117
  juncea, 108
Brassicaceae, 20, 116, 117
Brexiaceae, 7
Breynia, 148, 152, 154, 162
  cernua, 144
Bridelia, 152, 154
  minutiflora, 137
Bromus, 21
Broussonetia papyrifera, 151, 168, 169, 170, 171, 181
Brownlowia argentata, 34
Bruguiera, 32, 33, 34, 114, 116, 158, 175
  conjugata, 108
  eriopetala, 108
  gymnorhiza, 152

Brunoniaceae, 6
Bryophyllum, 137
Buchanania, 65
  arborescens, 160, 162
Buchnera ciliata, 147
  tomentosa, 75, 148
Buddlejaceae, 7
Bulbophyllum, 170
Burckella, 108, 124, 151, 175
  obovata, 124
Burseraceae, 86
Butomaceae, see Limnocharitaceae
Buxaceae, 6
Byblidaceae, 7

Cabombaceae, 6
Calamus, 143, 144, 146, 151, 152, 155, 156, 158, 159, 160, 162, 165, 170, 173, 174
  cf *hollrungii*, 142
Calanthe, 148, 167
Caldesia parnassifolia, 146, 147
Callicarpa, 140, 146, 147
  cana, 147
  caudata, 172
  pentandra, 158, 162
Callitrichaceae, 20
Callitriche, 20
Calophyllum, 46, 97, 152, 154, 157, 158, 159
  inophyllum, 29, 136, 139, 152, 155, 157, 158, 160, 162
Calycanthaceae, see Idiospermaceae
Calycacanthus magnusianus, 172
Camelliaceae, 7
Campanulaceae, 20; see also Pentaphragmataceae, Sphenocleaceae
Campnosperma, 42, 44, 45, 46, 49, 52, 158, 166
  brevipetiolata, 46, 174
  coriacea, 46
Camptostemon schultzii, 34
Cananga odorata, 51, 59
Canarium, 59, 64, 108, 116, 117, 122, 123, 148, 150, 158, 175, 177
  commune, see *C. indicum*, 108
  indicum, 81, 108, 123, 137
  mehenbethane, see *C. indicum*, 108
  salomonense, 108, 124
Canavalia maritima, 28
  obtusifolia, 148, 150
Capillipedium parviflorum, 57, 75, 148
Capparidaceae, see Emblingiaceae
Caprifoliaceae, 5, 20; see also Sphenocleaceae
Caprifoliaceae s.s., 6
Capsicum frutescens, 108, 114, 116, 117, 133
Carallia brachiata, 42
Cardamine, 20
Cardiopteris moluccana,· see *Peripterygium moluccanum*, 140
Cardiospermum halicacabum, 173, 174

Carex, 148
Carica papaya, 108, 115, 116, 117, 137, 140, 145
Carlemanniaceae, 5, 6
Carpha, 20
  alpina, 101
Carpodetus, 21, 87, 148, 152, 162
Cartonemataceae, 7
Caryophyllaceae, 20
Caryota, 114, 152, 155
  rumphiana, 108, 160, 162
Cassia, 109, 114
  alata, 61, 138
Cassytha filiformis, 28, 29, 165
Castanopsis acuminatissima, 65, 87, 88, 89, 97, 102, 109, 132, 147, 152, 155, 160, 162, 164, 177, 180
Castanospermum australe, 152, 154
Casuarina, 52, 82, 83, 103, 104, 105, 127, 128, 131, 148, 150, 152, 154, 155, 160, 180, 181
  cunninghamiana, 62, 83
  equisetifolia, 28, 29, 145, 162
  junghuhniana, 70
  oligodon, 162
  papuana, 68, 69, 74, 76, 78, 80, 83, 88, 97, 148, 162
Casuarinaceae, 18
Cayratia, 142
Cedrela toona, see *Toona ciliata*
Celastrus novoguineensis, 165
Celosia argentea, 109, 164, 172
Celtis, 65
Centella asiatica, 137, 138, 139, 140, 141, 145
Centrolepidaceae, 20
Centrolepis, 20
Cephalotaceae, 6
Cerastium, 20
Ceratophyllum, 38
Cerbera, 151
  manghas, 137, 138
  odollam, 157
Ceriops tagal, 32, 34
Chamaibainia, 22
Chenopodium, 136, 137
Chisocheton, 109, 132, 162
Chloranthaceae, 7, 20
Chloranthus, 109, 133, 148
  officinalis, 109
Choriceras tricorne, 53
Cinnamomum, 109, 133, 140, 142, 148, 150, 158, 162, 172
  culilawan, 144
Cissus, 137
Citriobatus, 14
Citrullus vulgaris, 115
Citrus, 109, 115, 116, 117, 124
Claoxylon, 143, 148, 152, 162
Cleistanthus, 148
Clematis, 143, 144
  clemensiae, 143
  glycinoides, 143

Clerodendron, 137, 144
  paniculatum, 172
Clerodendrum, see Clerodendron

Clethraceae, 7
Cochlospermaceae, 7
Cocos, 175
  nucifera, 108, 109, 122, 123, 135, 137, 145, 148, 152, 158, 169, 173, 175
Codiaeum variegatum, 137, 142, 145, 147, 148, 150, 164, 172
Coelachne, 21
Coelorhachis rottboellioides, 55, 75
Coix gigantea, 39, 109, 133, 134
  lachryma-jobi, 39, 109, 116, 117, 133, 151, 167, 170, 173, 174
Coleus, 109, 133, 139, 140, 143, 144, 147, 151, 172, 180
  atropurpureus, 148, 150, 172
  blumea, 172
  scutellarioides, 109, 142, 143, 148, 150, 164
Colocasia esculenta, 98, 107, 109, 115, 116, 117, 121, 125, 137, 143, 144, 145, 148, 175
Colona scabra, 148
Cominsia, 167
Commelina, 125, 129, 139, 175, 180
  cyanea, 109, 131
  diffusa, 109, 131, 142
Commelinacea, 146
Commelinaceae, 50, 174; *see also* Cartonemataceae
Commersonia, 79
  bartramia, 169
Compositae, *see* Asteraceae
Conandrium polyanthum, 162
Connarus, 9, 10
Convolvulaceae, 174
Convolvulus, 109, 120
Conyza, 19
Coprosma, 21, 99, 100, 102
Cordyline, 81, 127, 128, 139, 140, 142, 144, 162, 164, 170, 172, 177
  fruticosa, 136, 145, 148, 150, 164, 167, 172
  terminalis, 109
Coriaria, 9, 10, 20
Coriariaceae, 7, 20
Cornaceae, *see* Mastixiaceae
Corsia, 11, 14
Corsiaceae, 7
Corynocarpaceae, 7
Costus, 162
  speciosus, 139
Cotula, 19
Crassocephalum crepidioides, 137, 168, 180
Crassulaceae, 6
Crinum, 36, 148, 150
  asiaticum, 29, 157
  macrantherum, 139, 146, 172
Crossostylis, 14

Crotalaria, 76
  ferruginea, 148, 152
  linifolia, 109
  semperflorens, 167
Croton, 138, 148, 150, 162, 172
Crucicaryum, 20
Cruciferae, *see* Brassicaceae
Crypteroniaceae, 7
Cryptocarya, 64, 65, 81, 142, 148, 162, 172
Ctenitis, 109, 125
Ctenolophonaceae, 7
Cucumis, 139
  sativus, 109, 115, 116, 117
Cucurbita maxima, 109, 116, 117
  moschata, 109
  pepo, 109
Cucurbitaceae, 125, 174
Cunoniaceae, 7, 18, 20, 86
Cupressaceae, 7, 22
Curculigo orchioides, 56
Curcuma, 145, 146
  domestica, 109, 133, 151
Cyathea, 83, 89, 92, 102, 109, 125, 152, 155
  angiensis, 109, 167
  contamiana, 109
  rubiginosa, 109
  tenggerensis, 151
Cyatheaceae, 92
Cyathula prostrata, 147
Cycadaceae, *see* Zamiaceae
Cycas, 66, 125, 137, 138, 151, 177
  circinalis, 137, 146
  media, 76, 78
Cyclosorus, 38, 42, 138, 142, 143, 167
  truncatus, 109, 125
  unitus, 165
Cymbopogon, 140, 150
  citratus, 172
  flexuosus, 148, 150
  procerus, 75
Cynoglossum, 109, 173
  helwigii, 148
Cyperaceae, 20
Cyperus, 10
  melanospermus, 180
  pedunculatus, 28
  pedunculosus, 172
Cypholophus, 109, 133, 169
Cyphomandra betacea, 109
Cyrtandra, 11, 86, 109, 133, 148
Cyrtosperma, 116
  chamissonis, 107, 109, 114, 166

Dacrycarpus, 22, 91, 99, 102
  compactus, 100
Danthonia, 21, 91, 100
  archboldii, 91, 92, 105
  penicillata, 96
Daphniphyllaceae, 7, 20
Daphniphyllum, 20, 162
Davidsoniaceae, 6

Dawsonia, 86
Debregeasia, 165, 168, 169, 170, 171
Decaisnina hollrungii, 172
Decaspermum, 79, 109, 148, 150, 152, 162
Decatoca, 20
Degeneriaceae, 7
Dendrobium, 134, 143, 172, 173
Dendrocnide excelsa, 136, 160, 162
Dendromyza, 11
Dennstaedtia, 109, 125
Deplanchea tetraphylla, 76
Derris, 139, 146, 157
  trifoliata, 32
Deschampsia, 21
  klossii, 91, 92, 94, 96, 100, 105
Desmodium, 76, 133
  microphyllum, 109
  ormocarpoides, 66, 146
  repandum, 109
  sequax, 141
  umbellatum, 29, 78, 148, 151, 169
Detzneria, 21
Deyeuxia, 21, 91
Dianella ensifolia, 166
Dichelachne, 21
  novoguineensis, 91
Dicranopteris, 165
  linearis, 55, 80, 82, 83, 152, 170, 173
Dicrastylidaceae, 6
Dillenia, 162, 167
  alata, 48
Dimeria, 94
Dimorphanthera, 20, 151, 165, 173
Dioscorea, 107, 109, 115, 119, 125, 140, 143, 148, 175
  alata, 109, 115, 119, 178
  bulbifera, 109, 119, 178
  esculenta, 109, 115, 119, 178
  hispida, 109, 119, 178
  nummularia, 109, 119, 178
  pentaphylla, 109, 119, 178
Dioscoreaceae, *see* Petermanniaceae, Trichopodaceae
Diospyros, 30, 44, 50, 65, 109, 124, 132, 138, 152, 154
Diplazium, 109, 125, 133
  casperum, 109
  cordifolium, 109
Diplocaulobium, 173
Dipsacaceae, 20; *see also* Triplostegiaceae
Dipsacaceae s.s., 6
Dipterocarpaceae, 7, 52, 68, 73, 74, 86
Dischidia, 34, 78
Dodonaea viscosa, 128, 139, 150, 152, 154, 156, 160, 162, 180
Dolichandrone spathacea, 34, 138, 142
Dolichos lablab, 109, 116, 117, 125, 129, 130, 175
Donatiaceae, 6
Donax canniformis, 134, 162, 169
Dracaena, 172
Dracontonelum mangiferum, see Dracontomelon puberulum

Dracontomelon
  puberulum, 62, 81, 151, 153
Drapetes, 22, 101, 102
Drimys, 22, 86, 87, 100, 172
  piperita, 141, 145
Drosera, 58
Dryadodaphne, 20, 86, 97
Dryadorchis, 21
Drymaria cordata, 146, 147, 180
  rigidula, 162, 172
Dryopteris, 116, 117, 125, 133
  cf *arbuscula*, 109
  sparsa, 110
  truncata, 110
Dysoxylum, 65, 146, 160, 162
Dysphaniaceae, 6

Ebenaceae, 86
Ecdeiocoleaceae, 6
Echinochloa stagnina, 39
Echinopogon, 21
Elaeagnus, 165
Elaeocarpaceae, 20, 86, 87, 101
Elaeocarpus, 65, 97, 110, 132, 145, 148, 152, 153, 160, 162, 177
Elatostema, 50, 65, 86, 110, 133, 140, 141, 143, 145, 172
  blechnoides, 148
  cf *macrophyllum*, 133
Eleocharis, 48, 170
Eleusine indica, 136, 148
Ellisiophyllaceae, 7
Ellisiophyllum, 21
Elmerrillia papuana, 81, 84, 137, 148, 172
Embelia, 165
Emblingiaceae, 6
Emilia, 138
  prenanthoidea, 141
Endospermum, 59, 81, 84
  formicarum, 139, 140, 148, 150, 167
Engelhardtia, 97
Enhalus, 110, 114
Epacridaceae, 7, 20, 100, 101
Epiblastus, 21
Epilobium, 20
Epipremnum, 140
Equisetum, 100
  debile, 151, 153
Erechtites, 19
Eremosynaceae, 6
Eriachne, 55, 57, 58
Eriandra, 14
Ericaceae, 6, 20, 87, 99, 100
Erigeron linifolius, 146
  sumatrensis, 173
Eriocaulon, 48, 58
  australe, 133
Erythrina, 30, 66, 81, 110, 133, 146, 155, 158
  indica, see *E. orientalis*
  orientalis, 136, 153, 155

  variegata, 143, 144
Erythropalaceae, 6
Ethulia, 19
Eucalyptopsis, 74
Eucalyptus alba, 76, 78
  confertiflora, 76
  deglupta, 61, 62, 80, 83, 97
  papuana, 76
  tereticornis, 76, 78
Eucryphiaceae, 6
Eulalia leptostachys, 75, 91
Euodia, 59, 81, 97, 110, 114, 133, 140, 142, 143, 149, 150, 153, 154, 155, 156, 160, 162
  altata, 143
  anisodora, 142, 172
  bonwickii, 162
  crassiramis, 162
  elleryana, 158, 159, 172
  hortensis, 143, 172
Euonymus, 10
Euphorbia, 146
  buxoides, 146, 149, 150, 153, 155, 157
  hirta, 144
  plumerioides, 141, 157
  pulcherrima, 172
  serrulata, 75
Euphorbiaceae, 61, 174
Euphrasia, 10, 21
Eupomatia, 11, 13
Eupomatiaceae, 7
Euroschinus papuanus, 153, 155
Eurya, 22, 86, 162
Eurycles amboinensis, 140
Evodiella, 21, 153, 155, 156
  hooglandii, 148, 150
Excoecaria, 35, 139
  agallocha, 35
Fagaceae, 18, 20, 86, 175
Fagraea, 153
  ceilanica, 153
  racemosa, 167
Festuca, 21, 100
Ficus, 9, 19, 50, 52, 59, 61, 65, 66, 83, 110, 114, 116, 117, 119, 120, 125, 131, 134, 138, 139, 140, 141, 143, 145, 147, 148, 150, 151, 153, 154, 156, 158, 159, 160, 163, 166, 167, 168, 169, 177
  adenosperma, 162
  botryocarpa, 138
  botryocarpa var. *sub-albidoramea*, 110
  caulocarpa, 169
  copiosa, 110, 131, 153, 154
  dammaropsis, 110, 116, 117, 131, 156, 166, 167, 169
  glabella, 169
  gul, 152, 155
  gymnorygma, 168, 169
  iodotricha, 110, 168, 169
  itoana, 110
  microdictya, 169

  myriocarpa, 156, 158, 159
  nodosa, 110
  pachyrrhachis, 110, 138, 156, 168, 169
  pungens, 110, 140, 143
  robusta, 168, 169
  septica, 139, 140, 141, 142, 143, 145, 146
  trachypison, 153, 168, 169
  trichocerasa, 162
  wassa, 110, 131, 168, 169, 180
Fimbristylis, 28
Finschia chloroxantha, 110, 132
Flagellaria indica, 29, 42, 66, 145, 146, 147, 158, 159, 165
Flagellariaceae, see Joinvilleaceae
Fleurya, 139, 140
Flindersia, 52
Flindersiaceae, 7
Floscopa scandens, 110, 131
Frankeniaceae, 6
Freycinetia, 86, 165, 169, 172

Gahnia, 92
  sieberiana, 163
Gaimardia, 20
Galbulimima belgraveana, 149, 150
Galearia, 7
Galium, 21
Garcinia, 42, 44, 50, 65, 86, 94, 114, 124, 132, 150, 154, 166, 149, 150, 153, 154, 163, 166, 172
  teysmanniana, 172
Gardenia, 151, 153, 154
Garuga floribunda, 52, 66, 79
Gastrolepis, 20
Gaultheria, 20
Geitonoplesium cymosum, 165
Gentiana, 20, 92
Gentianaceae, 20
Geraniaceae, 20
Geranium, 20
Germainia capitata, 54, 57
Gesneriaceae, 7, 20, 86
Gibbsia, 22
Gironniera, 163
Giulianettia, 172
Gleichenia, 38, 55, 92, 96, 110, 133
  brassii, 165
  cf *hirta*, 172
  vulcanica, 100, 101
Glochidion, 41, 76, 151, 153, 163
  pomiferum, 163
Gmelina, 138, 158
  moluccana, 159, 167
Gnetaceae, 7
Gnetum gnemon, 65, 81, 110, 114, 116, 117, 119, 124, 131, 168, 169, 175, 177
Goodeniaceae, 7
Goodyera rubicunda, 110
Gordonia papuana, 163
Graptophyllum pictum, 110, 133, 164
Grevillea, 52, 54, 58

glauca, 53, 58
papuana, 76, 78, 149, 163
Grewia, 76
Gronophyllum chaunostachys, 110, 133, 153, 155
Gulubia, 94
Gunnera, 20
Gynostemma pentaphylla, 136
Gynotroches axillaris, 44
Gyrocarpus americanus, 66
Gyrostemonaceae, 6

Habenaria, 110
Halfordia, 52
Haloragaceae, 20
Haloragis, 180
Hamamelidaceae, *see* Altingiaceae, Rhodoleiaceae
Hanguana malayana, 38, 42, 43, 45
Harmsiopanax harmsii, 146
Hartleya, 20
Helicia microphylla, 163
  odorata, 169
  oreadum, 163
Heliconia, 10, 163, 164, 167
  microphylla, 163
  indica, 8
Heliconiaceae, 7, 8
Helicteres angustifolia, 53
Hemigraphis, 110, 131, 146, 147, 149, 150
Heritiera littoralis, 34
Hernandia nymphaeaefolia, 151
  peltata, *see H. nymphaeaefolia*
Heteromorpha arborescens, 146
Heteropogon contortus, 74, 75, 76
Hibbertia, 11, 13
Hibiscus, 136, 149, 150, 151, 154, 158, 159, 169
  manihot, 108, 110, 114, 116, 117, 119, 124, 129, 130, 139, 143, 144, 145, 146, 147, 175, 178
  rosa-sinensis, 147
  tiliaceus, 29, 35, 136, 138, 139, 143, 144, 156, 163, 168, 169
Hierochloë, 21
Himantandraceae, 7
Holochlamys, 149
Homalium foetidum, 153, 154, 158, 163
Homalomena, 172
Hopea, 69
  papuana, 69
Hornstedtia, 145, 146
  lycostoma, 110
Horsfieldia, 163
  spicata, 172
Hoya, 34, 83, 165
Hydnocarpus, 153, 154
Hydrangeaceae, 7
Hydrophyllaceae, 6
Hymenachne amplexicaulis, 39
Hypericum japonicum, 142
  macgregorii, 92

papuanum, 138
Hyptis, 136, 142

Icacinaceae, 20; *see also* Lophopyxidaceae
Idiospermaceae, 6, 8
Ilex, 86, 101, 153, 155
Illiciaceae, 6
Impatiens, 65, 110, 133, 136, 140, 141, 172
  hawkeri, 147
  linearifolia, 133
  mooreana, 133, 137
  nivea, 133
  platypetala, 133
Imperata, 145
  cylindrica, 30, 48, 53, 54, 55, 56, 74, 76, 82, 91, 92, 96, 97, 133, 134, 136, 137, 163, 164, 165
Indigofera, 76
Inocarpus edulis, 110, 116, 124
  fagiferus, 139
Intsia bijuga, 34, 51, 52, 62, 66, 150, 151, 163
Ipomoea, 147, 165
  aquatica, 39, 110, 116, 117
  batatas, 107, 110, 115, 116, 117, 122, 127, 137, 175
  congesta, 138
  pes-caprae, 28, 29, 138, 142, 145
  reptans, 110, 114
Iresine herbstii, 110, 151
Iridaceae, 7, 20
Isachne, 94
Ischaemum, 39, 55, 75, 91
  barbatum, 53, 54, 56, 57, 149
  muticum, 28, 29
  polystachyum, 39, 96, 163, 167
Ischnea, 19
Iteaceae, 6

Joinvillea, 14
Joinvilleaceae, 7
Juglandaceae, 7
Juncaceae, 20
Juncaginaceae, 7
Juncus, 20
Justicia, 172

Kaempferia galanga, 110, 133, 135, 146, 147, 172
Kalanchoe pinnatum, 138, 140
Kerigomnia, 21
Keysseria, 19
Kleinhovia hospita, 59, 134, 143, 144, 167
Koompassia, 74
Kopsia, 11
Korthalsella, 20

Lablab niger, *see Dolichos lablab*
Lactuca, 19, 110, 133, 138
  indica, 135, 151
  sativa, 110

Lagenaria siceraria, 110, 132, 167, 175
Lamiaceae, 180
Langsdorffia, 21
Laportea, 59, 61, 110, 133, 139, 140, 144, 169
  decumana, 145, 146, 147
  gigas, *see Dendrocnide excelsa*
Lauraceae, 18, 86, 156, 160, 175
Lecanthus, 22
Leea, 61, 163
  indica, 151
Leersia hexandra, 39, 94
Leguminosae, 20, 86, 174
Lemna, 37
Lepidagathis, 110, 131
Lepistemon ureceolatum, 165
Leucopogon, 101
Leucosyke, 22, 110, 133, 140, 156, 163, 168, 169
  candidissima, 140
Levieria, 163
Libertia, 20
Libocedrus, 22
Licuala, 50, 153
Liliaceae, 20, 175
Liliaceae, *see* Philesiaceae, Xanthorrhoeaceae
Limnocharitaceae, 6
Linaceae, *see* Ctenolophonaceae
Liparis, 170
Lipocarpha chinensis, 172
Lithocarpus, 65, 67, 87, 97, 163, 164
  cf *rufo-villosus*, 163
Litsea, 149, 172
Livistona, 41, 42
Loganiaceae, *see* Antoniaceae
Lophopyxidaceae, 7
Lophopyxis maingayi, 145
Loranthaceae, 20, 87
Lowiaceae, 6
Lucinaea, 110, 165
Ludwigia, 39
  adscendens, 146, 147
  octavalvis, 180
Lumnitzera racemosa, 34
Lunasia amara, 66, 167
Luzula, 20
Lycianthes, 149
Lycopersicon lycopersicum, 115
Lycopodium, 38, 55, 86, 110, 149, 172
  cernuum, 82
Lygodium, 55, 139, 151, 156, 166
  circinnatum, 158, 159, 165
Lysimachia, 21
  japonica, 110

Macaranga, 59, 116, 117, 124, 134, 138, 140, 141, 146, 158, 159, 163, 172
  aleuritoides, 146, 158, 163, 167
  tanarius, 158, 163, 167
Machaerina rubiginosa, 92, 96
Madhuca, 110, 124
Maesa, 139

edulis, 146
Magnoliaceae, 7
Mallotus paniculatus, 172
  philippinensis, 145, 147
  ricinoides, 145
Malvaceae, 175
Mangifera, 42, 52, 110, 116, 117, 124, 132, 158
  indica, 110, 132
  minor, 139
Manihot esculenta, 110, 115, 120
Manilkara, 30
Maniltoa, 50, 66, 153, 154
Maoutia, 168, 169, 171, 172
Marantaceae, 7, 50, 61
Maranthes corymbosa, 52, 151, 158, 159
Marattia, 139
Mastixiaceae, 7
Mearnsia, 20
  cordata, 149
Medinilla, 149, 165
Melaleuca, 30, 35, 42, 46, 47, 48, 49, 53, 54, 58, 74, 76
  cajuputi, 30, 42, 46, 54, 58
  leucadendron, 42, 54
  symphyocarpa, 53, 58
  viridiflora, 53, 54
Melastoma, 11
  affine, 180
  malabathricum, 76, 151
Melastomataceae, 7, 86, 175
Meliacea, 135
Meliaceae, 86, 175
Meliosma, 151, 163
Meliosmaceae, 7
Melochia tomentosa, 139, 140
Melothria, 173, 174
Memecylon, 153
Merremia peltata, 136
Messerschmidia, 29
  argentea, 172
Metroxylon, 107, 110, 112, 115, 116, 117, 140, 153, 159, 163, 165, 169, 175
  sagu, 43
Microcos, 50, 65, 110, 133
Microglossa pyrifolia, 138
Microlaena, 21
Microtatorchis, 21
Miscanthus floridulus, 92, 94, 96, 153, 155, 160, 163, 164
Mischocodon, 163
Mitragyna ciliata, 42
  speciosa, 41
Mitrastemon, 21
Monimiaceae, 7, 9, 18, 20; see also Atherospermataceae
Monostachya, 21
  oreoboloides, 100, 101
Monstera, 136
Montia, 21
Moraceae, 143, 155, 174, 178, 180
Morinda, 138

citrifolia, 110, 124, 136, 140, 142, 144, 145, 151
Mucuna, 145
Muehlenbeckia platyclada, 149
Muehlenbergia, 21
Musa, 98, 108, 110, 115, 116, 117, 120, 139, 140, 142, 144, 145, 147, 149, 158, 163, 167, 169, 173, 174, 175
Musaceae, see Heliconiaceae
Mussaenda, 153, 155
  ferruginea, 140, 142
  scratchleyi, 98
Myoporum, 11
Myosotis, 20
Myriactis, 19
Myricaceae, 7, 18
Myristica, 30, 50, 65, 110, 114, 116, 117, 146, 163, 166, 167
  hollrungii, 34, 46
  longipes, 172
Myristicaceae, 7
Myrmecodia, 34, 149
Myrsinaceae, 9, 86, 87, 99, 155
Myrtaceae, 9, 20, 55, 86, 87, 94, 99, 101, 175, 180

Nasturtium officinale, 110
Nastus, 86, 165
Nauclea, 42, 59, 76, 163
  coadunata, 41, 42, 44, 51, 56, 159
Neanotus, 21
Nelumbo nucifera, 38
Neonauclea, 66, 80, 153, 154, 163
Nepenthes, 9, 11, 42, 58, 149
  mirabilis, 98
Nephrolepis, 83, 150
  biserrata, 111, 118
Nertera, 11, 13, 21
  granadensis, 149
Nicotiana tabacum, 134, 138, 144, 145, 149, 181
Nothofagus, 2, 11, 13, 18, 20, 85, 87, 88, 89, 90, 97, 102, 104, 153, 154, 155, 160, 163, 180
  perryi, 94
Nymphaea, 38
Nymphaeaceae, 18; see also Barclayaceae, Cabombaceae
Nymphoides, 38
Nypa fruticans, 35, 111, 114, 135, 160, 163, 181
Nyssaceae, 6
Ocimum basilicum, 111, 133, 141, 142, 149, 150, 172
  sanctum, 139
Octamyrtus, 143
  arfakensis, 163
  pleiopetala, 153, 154
Octomeles sumatrana, 59, 60, 61, 62, 83, 149, 150, 159
Oenanthe javanica, 108, 111, 119, 125, 127, 128, 129, 131, 138, 140, 146, 175, 177, 180

Olacaceae, see Erythropalaceae
Olearia, 19, 87, 100, 153, 163
Omalanthus, 138, 149, 150, 167, 172
  novoguineensis, 163
  populneus, 163, 167
Onagraceae, 20
Oncosperma, 114, 153, 155
  tigillaria, 153
Oncothecaceae, 7
Ophiorrhiza, 111
Ophiuros tongcalingii, 53, 55, 75
Orania, 153, 155
Orchidaceae, 21, 175
Oreobolus, 13, 20
Oreocallis, 52
Oreocnide, 163, 164
Oreomyrrhis, 19
Oryza, 39
Osbeckia sinensis, 141
Oxalidaceae, 21; see also Averrhoaceae
Oxalis, 21, 149
  magellanica, 147
Oxychlamys, 20

Palaquium, 44
Palmae, 144, 150, 166, 174
Palmeria, 20, 111, 133, 163
Palmervandenbroeckia, 19
Pandaceae, 7
Pandanaceae, 177
Pandanus, 44, 58, 108, 111, 116, 117, 124, 126, 131, 132, 134, 137, 138, 149, 150, 151, 153, 155, 156, 157, 159, 160, 163, 164, 165, 166, 168, 169, 170, 173, 174, 175, 177, 181
  brosimos, 111, 132
  conoideus, 111, 126, 131, 151
  dubius, 29
  foveolatus, 111, 132
  julianettii, 111, 132
  papuanus, 157
  tectorius, 29
Pandorea pandorana, 165
Pangium edule, 81, 111, 124, 138, 149, 157
Panicum, 39
Papilionopsis, 20
Papuacedrus, 22, 91, 99, 133
  papuana, 88, 91, 111, 160, 164, 172
Papuapteris linearis, 100
Papuechites aambe, 138, 173
Papuzilla, 20
Paracryphiaceae, 7
Parahebe, 21
Parartocarpus venenosa, 111, 124, 138
Parietaria, 22
Parinari corymbosum, see *Maranthes corymbosa*
Parinari, see *Maranthes*
Parsonsia, 111, 143, 145
  pedunculata, 165
Paspalum, 136
  conjugatum, 136, 167, 180

Patersonia, 20
Pedilochilus, 21
Pemphis acidula, 29
Pennisetum, 138, 141
   macrostachyum, 61, 75, 153, 167, 173
Pentaphragmataceae, 7
Pentaphyllaceae, 6
Peperomia, 143, 146
Peracarpa, 20
Peripterygium moluccanum, 139
Perrottetia, 14, 15
Petermanniaceae, 6
Petrosaviaceae, 6
Phaius tankervilliae, 147
Phaleria macrocarpa, 168, 169
Phaseolus aureus, 115
   lunatus, 111, 116, 117, 119, 129, 130, 175
   vulgaris, 111, 115, 116, 117, 125
Phellinaceae, 7
Philadelphaceae, 6
Philesiaceae, 7
Philodendron, 140, 141
Philydrum lanuginosum, 48
Phragmites karka, 30, 32, 39, 41, 42, 43, 59, 61, 94, 163, 164
Phreatia, 173
Phrynium, 111, 163, 164, 167
Phyllanthus, 144, 145, 163
   amarus, 139, 140
   archboldianus, 163
   flaviflorus, 163
   nervosus, 153, 154
   urinaria, 142
Phyllocladus, 13, 22, 87, 91
Phytolaccaceae, 6
Phytophthora colocasiae, 122
Pilea, 111, 133, 149, 165
Pimelodendron amboinicum, 59, 65, 81
Pinaceae, 6
Piora, 19
Piper, 86, 111, 135, 139, 144, 145, 147, 149, 150, 153, 156, 157, 165, 173
   betle, 135
   methysticum, 135, 181
   stenocarpum, 111
Piptadenia novoguineensis, 146, 152, 153
Piptocalyx, 22
Pipturus, 83, 133, 146, 149, 151, 163, 166, 167, 168, 169, 173
   argenteus, 111, 141, 147
   repandus, 140, 141
   cf *verticillatus*, 169
Pisonia, 169
   longirostris, 156
Pistaciaceae, 6
Pistia stratiotes, 37, 38
Pithecellobium cf *sapindoides*, 173, 174
Pittosporum, 11, 86, 87, 100
   ferrugineum, 141
   pullifolium, 111, 132, 151
Planchonella, 86, 163, 173, 174

Planchonia papuana, 30, 51
Plantaginaceae, 7, 21
Plantago, 21, 100, 180
Plectranthus, 151, 164, 173
   scutellarioides, 138
Pluchea indica, 35
Plumbaginaceae, *see* Aegialitidaceae
Plumeria acutifolia, 139
Poa, 21, 100
   brassii, 163, 164
Poaceae, 21, 174
Podocarpaceae, 22, 180
Podocarpus, 91, 94, 99, 149, 153, 155, 159
   amarus, 163
   vitiensis, 163
Podostemonaceae, *see* Tristichaceae
Pollia, 111, 131
Polygala, 75
   paniculata, 147
Polygonum, 39, 180
   chinense, 111, 149
   minus, 149
Polyosma, 86, 164
Polypodium, 111, 133, 143, 144, 149, 150
   commutatum, 111
   irioides, 111
   linguaeforme, 111
Polyscias, 111, 133
   scutellaria, 147
Polystichum, 141
Pometia, 62, 64, 65, 84
   pinnata, 50, 124, 137, 164
Pongamia pinnata, 142, 145, 146, 157
Portulacaceae, 21
Posidoniaceae, 6
Potamogetonaceae, *see* Posidoniaceae, Ruppiaceae, Zosteraceae
Potentilla, 21, 92, 102
Pothos, 136
Pouzolzia, 133, 168
   hirta, 75, 111, 169
Premna, 173
   integrifolia, 138, 139, 141, 142, 143, 146, 151, 164
   obtusifolia, 159
Primula, 21
Primulaceae, 21
Procris, 111
Prosopis insularum, *see Piptadenia novoguineensis*
Proteaceae, 7-18
Protium macgregorii, 66
Prunus, 86, 171
Pseudopogonatherum irritans, 75
Pseudoraphis spinescens, 39, 41, 48
Psidium guajava, 111, 116, 117, 139, 145
Psophocarpus tetragonolobus, 111, 116, 117, 119, 125, 129, 175
Psychotria, 139, 141, 149
Pteridium, 55
   aquilinum, 141, 147

   cf *aquilinum*, 149
Pteris, 167
   moluccana, 111, 116, 133
Pterocarpus indicus, 30, 52, 62, 63, 66, 138, 139, 141, 144, 145, 146, 151, 164
Pterostylis, 21
Pueraria, 76, 165
   lobata, 111, 132, 177
   phaseoloides, 135, 147
Pullea, 20
   glabra, 153, 173
Pygmaeopremna sessilifolia, 56
Pyrolaceae, 7, 8

Quassia, 145
Quintinia, 11, 13, 21, 101, 153, 154, 164

Rafflesiaceae, 21
Ranunculaceae, 21
Ranunculus, 21, 92, 102
   pseudolowii, 111, 149
Rapanea, 100, 102, 149, 153, 154, 164, 173
Remirea maritima, *see Cyperus pedunculatus*
Restionaceae, 7; *see also* Ecdeiocoleaceae
Rhamnus nepalensis, 149, 165
Rhaphidophora, 136
Rhipogonum, 142
Rhizophora, 32, 33, 34, 164
Rhizophoraceae, 151; *see also* Anisophylleaceae
Rhodamnia, 52
Rhododendron, 10, 11, 12, 13, 14, 100, 146, 149, 151, 173
   gracilentum, 111
Rhodoleiaceae, 6
Rhodomyrtus, 133
   novoguinensis, 111
Rhus, 30, 79, 153, 164
   taitensis, 83, 153, 155, 156
Rhynchospora rugosa, 96
Rhyticaryum, 145
Riedelia, 173
   carallina, 111, 167
   monticola, 173
Rorippa, 111, 129, 131, 180
Rosaceae, 21; *see also* Stylobasiaceae
Rubiaceae, 21, 86, 99, 155
Rubus, 86, 132, 133, 136, 137
   cf *fraxinifolius*, 111
   glomeratus, 141
   ledermannii, 147
   moluccanus, 111, 134, 147, 165, 166
   rosifolius, 111, 139, 149, 180
Rungia klossii, 108, 111, 127, 128, 129, 130, 131, 175, 177
Ruppiaceae, 6
Rutaceae, 18, 21, 175; *see also* Flindersiaceae

Sabiaceae, 7; *see also* Meliosmaceae

Saccharum, 155, 164
   edule, 108, 111, 116, 117, 119, 125, 129, 175
   officinarum, 108, 111, 116, 117, 125, 143, 144, 145, 149, 175, 177
   robustum, 39, 40, 41, 59, 61
   spontaneum, 55, 56, 61, 75, 82, 139
Sacciolepis indica, 138
Sagina, 20
Salicaceae, 6
Salvadoraceae, 6
Samanea saman, 146
Sambucus, 20
Santalaceae, 87
Sapindacea, 155
Sapium indicum, 34
Sarcospermataceae, 7
Saurauia, 86, 111, 133, 164, 167, 173
Saururaceae, 6
Saxifragaceae, 21; see also Baueraceae, Eremosynaceae, Hydrangeaceae, Iteaceae, Philadelphaceae, Tetracarpaeaceae
Saxifragaceae s.s., 7
Scaevola, 29, 142, 145
   frutescens, see *S. taccada*
   oppositifolia, 142, 165
   taccada, 138, 142, 143, 146, 147
Schefflera, 99, 145, 164, 167, 173
Schisandraceae, 6
Schismatoglottis, 10
Schizomeria, 164, 165
Schoenus, 55, 57, 58
Schuurmansia, 87
   henningsii, 154
Scirpus mucronatus, 173
Scleranthus, 20
Scleria, 38, 39, 111, 133, 166
   chinensis, 153, 154
Scrophulariaceae, 21; see also Ellisiophyllaceae
Scyphostegiaceae, 6
Securinega virosa, 142
Sehima nervosum, 75
Selaginella, 50, 58
   caudata, 153, 166
   opaca, 111
Semecarpus, 139
Senecio, 20
Sepalosiphon, 21
Sericolea, 11, 13, 20, 86
Sesuvium portulacastrum, 34, 35
Setaria palmifolia, 108, 111, 116, 117, 119, 125, 127, 128, 129, 130, 131, 170, 175, 177
Shorea, 11
Sida acuta, 145
   acuta ssp. *carpinifolia*, 145
   cordifolia, 141, 145
   rhombifolia, 136, 137, 141, 147
Simarouba, see *Quassia*
Sinoga lysicephala, 52, 55, 58
Sloanea, 65, 97, 164

   archboldiana, 111, 132
Smilax, 142
Smithia sensitiva, 164
Solanum, 10
   melongera, 111
   nigrum, 108, 111, 129, 131, 175, 180
   torvoideum, 149
   tuberosum, 111
Sonchus, 111
Sonneratia, 32
   alba, 164
   caseolaris, 32
Sopubia trifida, 149
Sorghum nitidum, 55, 75
Sparganium, 21
Sparganiaceae, 21
Spathodea, 81
Spathoglottis, 134, 149, 166, 167
Sphagnum, 86
Sphenocleaceae, 6
Sphenostemon, 21
Sphenostemonaceae, 7, 21
Spiraeopsis, 20, 164
   celebica, 164
Spiranthes, 21
Spondias dulcis, 112, 124, 149
Sporobolus virginicus, 35
Stachytarpheta, 96
Stackhousia, 11
Stapelieae, 3
Staphyleaceae, 7
Stellaria, 20, 112, 149
Stenochlaena, 29, 42, 112
Stephania, 142
Sterculia, 59, 66, 81, 84, 112, 132, 164, 169, 177
Sterculiaceae, 175
Strasburgeriaceae, 8
Streblus, 169
   urophyllus, 164
Stylidiaceae, 7; see also Donatiaceae
Stylobasiaceae, 6
Styphelia, 102
   suaveolens, 92, 105
Styracaceae, 7
Swertia, 20
Symbegonia, 112, 143, 145
Symphoremaceae, 6, 8
Symplocos, 86, 151, 153, 154, 164
Syzygium, 30, 42, 44, 52, 65, 66, 94, 97, 112, 124, 132, 138, 139, 143, 144, 145, 149, 153, 154, 156, 164, 165, 167, 168, 169, 177
   aqueum, 112, 124
   malaccense, 112, 114, 116, 117, 124
   sp. aff. *pachyclada*, 112

Tagetes, 151
Tapeinocheilos dahlii, 139
Taxaceae, 6
Taxodiaceae, 6
Tecticornia cinerea, 34
Tephrosia, 76
Terminalia, 30, 45, 50, 52, 59, 64, 66,

   108, 175, 177
   brassii, 46, 47, 49
   canaliculata, 44
   catappa, 29, 112, 114, 124, 166
   complanata, 51
   copelandii, 112
   impediens, 112
   kaernbachii, 81, 112, 124
   solomonensis, 112, 124
Ternstroemia, 146, 164, 173
   cherryi, 138
Tetracarpaeaceae, 6
Tetracera, 147
Tetragoniaceae, 6
Tetrameristaceae, 6
Tetramolopium, 19, 20
Teysmanniodendron bogoriense, 51
Thalictrum, 21
Theacea, 157
Theaceae, 7, 22; see also Bonettiaceae, Camelliaceae, Tetrameristaceae
Thelymitra, 21
Thelypteris, 112, 133
Themeda australis, 39, 54, 55, 56, 57, 74, 75, 76, 78
Thespesia peekeli, 170
   populnea, 159
Thoracostachyum sumatranum, 38
Thymelacaceae, 22
Timonius, 42, 61, 76, 82, 83, 141
   avensis, 164
   timon, 78, 139, 142
Toona ciliata, 159
Tournefortia sarmentosa, 142, 145; see also *Messerschmidia*
Trachymene, 19
Trapaceae, 6
Trema, 61, 82, 83, 103, 128, 149, 150, 156, 164, 169, 180
Tremandraceae, 6
Trichopodaceae, 6
Trichosanthes, 112, 116, 117, 132
Tridax procumbens, 138
Trigoniaceae, 6
Trigonotis, 20
   procumbens, 112
Trimenia, 22, 145, 149, 164
   papuana, 164
Trimeniaceae, 7, 22
Triplostegia, 20
Triplostegiaceae, 7
Tripogon, 21
Trisetum, 21
Tristania, 43, 52, 53, 54
   suaveolens, 58
Tristichaceae, 6
Triumfetta, 146
   nigricans, 133
   pilosa, 165, 168, 169
   rhomboidea, 149, 165
Trochocarpa, 20, 101
Typha, 153
Umbelliferae, see Apiaceae

Uncaria, 112, 165
Uncinia, 20
Urena lobata, 169
Urtica, 22
Urticaceae, 7, 22, 168, 174, 177, 178, 180
Usnea, 78, 91
Utricularia, 37, 58

Vaccinium, 100, 101, 164, 173
   acrobracteatum, 149
   cf *auriculifolium*, 165
Valerianaceae, 6
Vatica papuana, 67, 69, 136, 137
Verbenaceae, *see* Dicrastylidaceae, Symphoremaceae
Vernonia, 146, 173
   lanceolata, 149
Veronica, 21
Villebrunea rubescens, 142
Viola, 22, 138, 168, 180
   betonicifolia, 112
   klossii, 141
Violaceae, 22
Vitex cofassus, 51, 150, 153, 154, 164

Wahlenbergia marginata, 112, 149
Wedelia biflora, 136, 137, 141, 143, 145
Weinmannia, 14
Wendlandia, 140
   paniculata, 139, 141, 164, 167
Wikstroemia androsaemifolia 169
Winteraceae, 2, 19, 22

Xanthomyrtus, 20, 99, 153, 154, 164
Xanthorrhoeaceae, 7
Xanthosoma sagittifolium, 107, 112, 114, 121, 125, 129, 167
Xanthostemon, 43, 52, 53
Xylocarpus moluccanus, 34

Zamiaceae, 6
Zannichelliaceae, 6
Zea mays, 112, 116, 117
Zingiber, 112, 127, 133, 136, 137, 138, 139, 145, 146, 149, 150, 173
   officinale, 133
   zerumbet, 112, 137, 141, 143, 144
Zingiberaceae, 133, 175
Zosteraceae, 6

General Index

Abelam, 119
Admiralty Islands, 134, 157
adventitious roots, 34, 35, 41, 42, 51, 65, 86
adzes, 152
aerial roots, 34
Aibika, *see Hibiscus manihot*
Aitape, 159
algae, 58
alluvium forest, 34, 37, 46, 49-51, 64, 65, 70
altitude, effect of, 65, 66, 86-8
ant house plant, 34
aquatic, 114
aquatic vegetation, 37-8, 42, 48, 66
Arfak, 2
arrows, 152-5, 156
ash-salt, 133-4, 141
association, of species, 73, 74, 96, 105
axes, 152-3

Baliem, 126, 128, 137, 150
balsams, *see Impatiens platypetala*
bamboo, 50, 52, 59, 64, 66, 80, 81, 83, 84, 85, 86, 88, 94, 99, 112, 114, 125, 133, 134, 150, 151, 152, 154, 155, 156, 170
banana, 36, 59, 106, 107-8, 114, 115, 118, 119, 120, 121, 122, 126, 127, 129, 144, 146, 150, 175, 177, 178, 179, 181, 182; yield, 120; nutritional value, 115, 120; wild, 80; *aee also Musa*
bandages, 139
Barringtonia formation, 29
beans, 114, 115, 119, 127, 129
beard moss, *see Usnea*
beech, *see Nothofagus*
betelnut (palm), 80, 135, 141; *see also Areca catechu*
betel-pepper, 135
Biak Island, 112, 121
bilas, *see* personal ornaments
bilum, *see* string bag
bird lime, 156
bird whistle, 156
black spear grass, *see Heteropogon contortus*
Bosavi, 137, 139, 142
Bougainville, 46, 83, 94, 122
bows, 152-5
breadfruit, 59, 119, 123, 132, 138, 141, 177, 178, 181; nutritional value, 123; wild, 114; *see also Artocarpus altilis*
breathing roots, *see* pneumatophores
bronchitis, 144
Bulolo, 84
Bulolo River, 64
bunch spear grass, *see Heteropogon contortus*
burning, *see* fire
burns, 137
buttresses, 34, 46, 50, 59, 64, 86

candle nut oil, 174
canoes, 157-9
Cape Vogel Peninsula, 74
carvings, 150-1
cash crops, 183
cassava, 36, 115, 118, 119, 120; yield, 120; nutritional value, 115, 120; *see also Manihot esculenta*
Central District, 139, 144, 146
centre of diversity, 3
ceremonial, 118, 119, 126, 166, 172-3, 174, 183
chance dispersal, 74
chance establishment, *see* chance dispersal
chili, *see Capsicum frutescens*
Chimbu (District), 126, 128, 151, 181, 183
citrus, 118
climate, effect of, 25, 31, 52, 66, 86-8, 102-4, 105
climax forest, *see* climax vegetation
climax vegetation, 59, 62, 74, 79-81, 104
climbers, 29, 32, 41, 42, 45, 51, 52, 59, 64, 78, 80, 86, 96
climbing ferns, 42, 45, 51, 64
cloth, 168-70
clothing, 170
cloud cover, effect of, 65, 86-7
club moss, *see Lycopodium cernuum*
clubs, 121, 152-3, 154, 155
coconut, 30, 36, 108, 113, 114, 118, 119, 122-3, 146, 150, 151, 170, 177, 181, 183; nutritional value, 122, 123
colds, 143-4
condiments, 114, 118, 131, 133-4, 150
conifers, 86, 87, 88, 91, 94, 96, 97, 99, 100, 101, 104
coniferous forest, *see* conifers

conservation, 37, 62
continental drift, 17, 19
contraceptives, 146, 147
coral islands, 28, 30
coughs, 143-4
creepers, 28, 29, 62, 80, 96
cucumber, *see Cucumis sativus*
cultivars, 106, 119
cushion herbs, 99, 100
cuts, 136-7

deciduous trees, 52, 66
demarcation lines, 1
diarrhoea, 144-6
digging sticks, 118, 121, 152-3, 154
Digoel River, 112
dipterocarps, *see* Dipterocarpaceae
disjunct distribution, 3
dispersal ability, 4, 9
dispersal, long distance, 4
drainage, effect of, 25, 31, 37, 51, 53, 57, 58, 61, 100
dressings, 139
drills, 154
drought, effect of, 129
dry evergreen forest, 52-3, 58, 104
dysentery, 144-6

ear infections, 142
Eastern Highlands, 126, 127, 128, 133, 170
ecological amplitude, 3, 27, 68, 94
ecological tolerance, 3, 4, 9, 75, 91, 96
edaphic climax, 42, 48, 57, 58, 84
Eilanden River, 112
elfin woodland, 88
endemic families, 7
endemic genera, 14
endemics, 3
endemism, 8, 18, 19, 101
Enga (District), 126, 127, 134, 183
d'Entrecasteaux Islands, 2, 139, 142, 146
environmental tolerance, *see* ecological tolerance
epiphytes, 32, 52, 65, 66, 78, 80, 87, 94, 100
epiphytic climbers, 51; ferns, 51, 86; mosses, 86, 94; orchids, 51, 87-8
Erima, *see Octomeles sumatrana*
eucalypts, 53, 55, 74, 78
euphorbs, *see* Euphorbiaceae

eye infections, 142

ferns, 29, 32, 37, 38, 51, 59, 62, 65, 66, 78, 79, 80, 82, 83, 86, 94, 98, 100, 114, 125, 178
fevers, 142-3
fig trees, 125, 138
finger fern, see *Papuapteris linearis*
Finisterre Ranges, 133
fire, effect of, 38, 42, 48, 52-3, 55, 57, 58, 69-70, 75, 76, 79, 91, 92, 95-6, 97
fishing tools, 156-7
flavouring, 133-4
flooding, see inundation
Flora Malesiana, 4, 8, 24
floristic richness, 5
Fly River, 37, 41, 42, 104, 112, 121, 159, 178
food crops, new, 183
food containers, 166-7
food plants, 108-12, 175
food preparation, 166-7
Fore, 133, 135
fossils, 4
Frederik-Hendrik Island, 115, 118, 135, 182
french bean, see *Phaseolus vulgaris*
fresh water mangrove, 46
frost, effect of, 91, 94-5, 96, 129
fruit, edible, 80, 108-12, 114, 118, 126, 127; nutritional value, 116-17, 123
fruit trees, 118, 119, 122-4, 131-2, 178
fungi, 114

gallery forest, 76, 79
Gazelle Peninsula, 183
Geelvink Bay, 159
ginger, 29, 50, 52, 59, 61, 76, 79, 80, 86, 141, 150; see also *Zingiber zerumbet*
Gondwanaland, 17
gourds, 132, 134, 167-8, 170, 176
grass, edible, 108, 125, 129; nutritional value, 125; see also *Setaria palmifolia*, *Saccharum edule*
grass swamp, 38-42, 48
green vegetables, 108, 114, 118, 119, 122, 124-5, 126, 127, 129, 130-1, 166; yield, 130; nutritional value, 124
guava, 124
Gulf District, 112, 135, 158, 168

hanging gardens, 114
heart 'cabbage', 114, 123, 133
heath, 55, 92, 100
herbaceous beach vegetation, 28-9
herbaceous fern vegetation, 55
herbaceous swamp vegetation, 37, 38-9, 48, 70, 97-8, 100-1

Highlands Districts, 85, 97
Hood Bay, 64
hoop pine, see *Araucaria cunninghamii*
horsetails, see *Equisetum*
houses, 160-6
human activity, see man-made vegetation
Huon peninsula, 122
hyacinth bean, 125, 129-30; nutritional value, 130; see also *Dolichos lablab*

internal haemorrhage, 146
inundation, effect of, 29-30, 35, 37, 39, 42-3, 46, 48, 49, 51, 59-62
Irian Jaya, 52, 90, 104-5, 112, 115, 118, 126, 128, 133, 135, 136, 152, 156, 159

Jaya Mountains, see Mt Jaya
Jimi Valley, 131, 133, 139, 155, 156

Kaironk, 132
Kamarere, see *Eucalyptus deglupta*
kangaroo grass, see *Themeda australis*
kapok, 80
Karuka, see *Pandanus brosimos*
kauri pine, see *Agathis macrophylla*
kava, see *Piper methysticum*
Keraki, 135
Kerema, 64
Kikori River, 112
Kiwai, 135
Klinki pine, see *Araucaria hunsteinii*
knee roots, 34, 46
knives, 152-3, 154
Kukukuku, 126, 133, 135, 155, 168
kunai, see *Imperata cylindrica*
Kutubu, 167, 168
Kwila, see *Intsia bijuga*

Lae, 84
Lake Inim, 181
land connections, 4, 19
land slides, 81-2
laurel family, see Lauraceae
leaf litter, 51, 52
legumes, 75, 91, 120
lianes, 45, 51, 59, 64, 66, 86, 87, 100
lichens, 95, 99, 100
light, effect of, 32, 41, 43, 44-5, 50, 56, 70, 76, 80, 131
lightning damage, 32-3, 79
lima bean, see *Phaseolus lunatus*
lime, 135, 137, 138, 139, 141, 142
limestone, 68, 69, 88, 89, 91
limit of tree growth, see treeline
littoral forest, 29-30
lower montane zone, limits, 25
lowlands, upper limit, 25

Madang, 64
magic, 118, 136, 146, 147-50, 181

malay apples, 114
Mamberamo River, 112
Mandobo, 112
mangoes, 124
mangroves, 16, 31-7, 104, 106, 114, 155, 177
man-made vegetation, 27, 30-1, 36, 59, 80-1, 96-7, 102, 103
Manus, 139, 142, 144
Maprik, 183
Marind(-Anim), 112, 135
marita, see *Pandanus conoideus*
Markham River, 56, 64, 91, 120, 177
masks, 150-1
Mendi, 147,
Menyamya, 133
Merauke, 52, 104
milkwood, see *Excoecaria agallocha*
Milne Bay District, 154
montane flora, 14-15, 18
Morobe, 135
mosses, 65, 85, 86, 94, 99, 100
moss forest, 87
mounds: crab, 36; for planting, 127; termite, 58
Mt Albert Edward, 95, 99, 105
Mt Bangeta, see Saruwaged Range
Mt Doorman, 104
Mt Giluwe, 96, 99
Mt Hagen, 101, 139, 142, 144, 154, 155, 179, 180
Mt Jaya, 99, 104, 105
Mt Mandala, 105
Mt Suckling, 74
Mt Trikora, 96, 99, 100, 104, 105
Mt Victoria, 99
Mt Victory, 79
Mt Wilhelm, 95, 96, 99, 100, 101, 102, 103, 105
Musa River, 64
musical instruments, 151-2
mycorrhiza, 90

Neon Basin, 95
nets, 155-6
New Britain, 61, 62, 65, 89, 121, 135, 137, 139, 142, 154, 156, 158, 159, 160, 170, 183
New Guinea Rosewood, see *Pterocarpus indicus*
New Guinea Walnut, see *Dracontomelon puberulum*
New Ireland, 158, 159
nipa palm, 34, 35-6, 37, 106, 113, 114; see also *Nypa fruticans*
Nissan Islands, 159
Northern District, 139, 158
nuts, edible, 108, 114, 118, 122, 135; nutritional value, 116-17
nut trees, 119, 122-4, 131-2, 178

oak, see *Lithocarpus*
oil palm, 62

Okapa, 133
orchids, 34, 51, 83, 86, 87-8, 134
ornamental species, 81
Orokaiva, 134, 139, 151, 170, 174

pains, general body, 139-41
paintings, 150-3
palm forest, 94
palms, 29, 34, 42, 50, 52, 64, 80, 86, 135, 177, 181; *see also* Palmae
pandans, 30, 34, 38, 41, 42, 43, 44, 50, 51, 56, 59, 61, 80, 86, 94, 99; *see also Pandanus*
papaya, 115, 118, 119, 124, 141
paper-bark, *see Melaleuca*
Patep, 126
penis cases, 170
personal ornaments, 170-4
pes-caprae formation, 28, 29
pineapples, 36, 115, 124
pioneer plants, 28, 29, 31, 32, 34, 59, 61, 70, 95, 99
pitcher plants, *see Nepenthes*
pit-pit, 55, 119; edible, 126, 129
plant-ash, 133-4
plant families: distribution of, 8; number of, 5-8
plant genera: distribution of, 9-16; distribution types of, 9-11; montane, 14-16, 19-22; number of, 11-14
plant species, number of, 16
plant succession, 48-9, 59-62, 80, 81-4, 96-7
plate tectonics, 17, 19
pneumatophores, 31, 34, 43, 46
poisons, 146
Popondetta, 64, 78
population density, 118-19, 121, 126, 183
Port Moresby, 2, 14, 66, 120
profile diagrams, 72-3
pumpkin, 125
Purari River, 112

rafts, 157-9
Rai coast, 122, 124, 135
rainfall, effect of, 31, 34, 52, 55, 66-7, 75, 76
Ramu River, 42, 56, 91, 112, 155, 177
range of habitat, *see* ecological amplitude
rattan, 46, 51, 52, 59, 64, 80, 86, 155, 156; *see also Calamus*
relic, 70, 90
relic distribution, 3
reservation, *see* conservation
respiratory complaints, 144
rheophytes, 66
rice grass, *see Leersia hexandra*
ritual, 126, 136, 146, 147-50, 151, 172, 181
rock type, effect of, 67-8, 88, 99

rosette herbs, 99, 100
rubber, 62

sago grubs, 137
sago palm, 27, 30, 38, 42, 43-4, 44, 46, 49, 51, 61, 106, 112-15, 126, 151, 177, 178, 181, 182; yield, 114; nutritional value, 114, 115; *see also Metroxylon sagu*
salt couch grass, *see Sporobolus virginicus*
salt flats, 31, 34
salt spray, effect of, 29
sand couch, *see Sporobolus virginicus*
Saruwaged Range, 99
savanna, 52, 66, 106, 176, 177, 178; *Casuarina papuana*, 74; eucalypt, 25, 29, 35, 64, 69, 74, 76-8, 79, 104; *Melaleuca*, 54-5, 74; mixed, 53-4; tree fern, 92, 96, 105
Schrader Mountains, 132
scrub, 31, 66; beach, 29; lower montane zone, 88, 95, 96; mangrove, 32, 34-5; *Sinoga*, 55; upper montane zone, 98, 99, 105
seaweed, 114
secondary forest, 59, 69, 81, 84, 97
sedatives, 146, 147
sedge-grassland, 53, 55, 57-8
sedge-grass swamp, 91, 92-4, 95
sedges, 28, 34, 35, 37, 38, 42, 43, 45, 51, 55, 57-8, 61, 66, 92, 98, 100
seeds, edible, 114, 130, 132
Sepik River, 37, 39, 42, 44, 55, 56, 79, 112, 119, 155, 159, 177
shade, effect of, *see* light
shifting cultivation, 57, 64, 81, 92, 118-19, 126, 180; upper limit, 85, 94
sieva beans, *see Phaseolus lunatus*
skin diseases, 138-9
snares, 156
soil, effect of, 31, 51-2, 67-8, 74, 88, 100
soil salinity, effect of, 34, 35
sores, 138-9
sore throats, 143-4
soursop, 118
southern beech, *see Nothofagus*
Southern Highlands, 89, 92, 101, 126, 127, 128, 137, 139, 142
spades, 152-4
spears, 152-6
species/area curves, 71, 73-4
species diversity, 71, 73-4
spinach, 125; yield, 131; nutritional value 125; *see also Amaranthus, Rungia klossii*
staple food, 107, 118-22, 128-9
Star Mountains, 121, 174
stilt roots, 31, 34, 46, 50, 51, 86
stimulants, 147, 150
stomachaches, 144-6
Strickland River, 42, 121, 178

string, 168-9, 170
string bag, 166
subsidiary crops, 132-3
suckers, 43, 90, 120
sugar cane, 36, 108, 114, 115, 119, 125-6, 127, 129, 144, 150, 178, 179; *see also Saccharum officinarum*
sugar palm, *see Arenga pinnata*
sundew, *see Drosera*
supplementary crops, 108, 115, 118, 129-31, 182
supplementary food, *see* supplementary crops
swales: littoral, 30, 42; riverine, 61
swamp cultivation, 115, 118
swamp forest, 34, 37, 51, 89; *Campnosperma*, 46; lower montane, 91, 94, 95; *Melaleuca*, 42, 46-8; mixed, 44-6; *Terminalia brassii*, 46
swamp grass, 30, 34, 37, 38, 42, 48, 59, 61
swamp savanna, 37; *Melaleuca*, 42; mixed, 42
swamp woodland, 30, 34, 37; mangrove, 34-5; mixed, 42-3; nipa, 35-6; pandan, 44; sago, 43-4
sweet potato, 36, 107, 115, 118, 122, 126, 128-9, 137, 179, 180, 182, 183; yield, 122, 129; nutritional value, 115, 122; *see also Ipomoea batatas*
swidden gardening, *see* shifting cultivation
sword grass, *see Miscanthus floridulus*

Table Bay, 62
Tahitian Chestnut, *see Inocarpus edulis*
Tari, 137, 142, 150, 154
taro, 107, 114, 115, 118, 119, 121-2, 124, 126, 127, 129, 137, 150, 178, 179, 180, 182, 183; yield, 121; nutritional value, 115, 122; *see also Colocasia esculenta*
Taun, *see Pometia pinnata*
Tauri River, 64
Telefomin, 126, 133
tidal regime, effect of, 31, 32, 34, 35
tigasso oil, 166, 174
ti-tree, *see Melaleuca*
tobacco, 118, 134, 150, 181; *see also Nicotiana tabacum*
toddy, 123, 135
tomatoes, *see Lycopersicon lycopersicum*
toothache, 141
topography, effect of, 25, 31, 65-6, 88
Tor, 112
Torres Strait, 1, 13, 14, 16, 17, 176, 178
Trans-Fly, 119, 135, 152, 167
traps, 156
tree density, 70-2
tree ferns, 50, 64, 80, 83, 86, 92, 94, 96, 99, 102, 103
tree limit, *see* treeline
treeline, 95, 98, 100, 102, 103, 104, 105,

177, 178, 179
Trobriands, 119, 142, 144, 157, 183
tuberculosis, 144
Tugeri, 135

ultrabasic rocks, 68, 88

Vanimo, 62
vertical stratification, 72-3
Vogelkop, 121, 126, 135
volcanic activity, effect of, 82-4

Wahgi valley, 127, 179
Wallace line, 1
Waropen, 112, 114, 135, 156, 159
water couch grass, *see Pseudoraphis spinescens*
watermelon, *see Citrullus vulgaris*
water plants, *see* aquatic vegetation
water salinity, effect of, 31-2
Watom Island, 158
Western District, 158
Western Highlands, 92, 101, 126, 127, 128
wild nutmeg, *see Myristica*
wild rice, *see Oryza*
wild strawberry, *see Rubus* sp.
winged bean, 125, 129-30; yield, 130; nutritional value, 125, 130; *see also Psophocarpus tetragonolobus*
Wissel Lakes, 126, 128, 151, 159
woodland, 66; beach, 29; upper montane, 98
wounds, 136-7

yams, 106, 107, 114, 115, 118, 119, 120, 127, 129, 150, 176, 179, 180, 182, 183; yield, 119, 183; nutritional value, 115, 119; wild, 106, 125, 126; *see also Dioscorea*

Text set in 10 point Times and printed on 115 GSM Matt Art by Colorcraft Ltd., Hong Kong